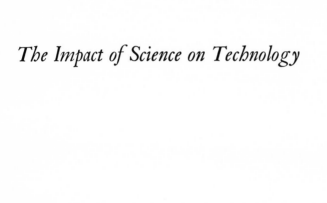

The Impact of Science on Technology

THE
IMPACT OF SCIENCE
ON TECHNOLOGY

Edited for the Columbia University Seminar
on Technology and Social Change by
AARON W. WARNER, DEAN MORSE,
and ALFRED S. EICHNER

Columbia University Press   New York and London
1965

Preface

It is important to keep in mind the nature of the papers included in this volume. Actually, the term "papers" is something of a misnomer, for these are not papers in the formal sense. More accurately, they are informal talks that were delivered before the Columbia University Seminar on Technology and Social Change. It is this fact which accounts for the volume's somewhat colloquial tone. As editors, we have not attempted to expunge that tone, for we have felt that the important ideas contained in these talks cannot be better expressed than in the authors' own words. It goes without further elaboration that whatever merit this volume has derives from the thoughts and reflections of these men, most of whom will be familiar to the reader as eminent scientists and noted public officials. We, as editors, are in their debt not only because they have provided the very substance of this book but also because they have been gracious and kind enough to help us in preparing their manuscripts for publication.

As these were talks delivered before a seminar, they were followed, in all cases, by a lengthy and vigorous discussion. Often this discussion gave the speaker a chance to elaborate on or sharpen points made in his earlier remarks. Often, too, it gave the members of the seminar itself a chance to present or test opposing points of view. Again, as editors, we have attempted to preserve the integrity of this dialogue. At the same time, however, we have tried to eliminate all repetitive or extraneous comments. How well we have reconciled these conflicting objectives the reader will have to judge for himself. As a practical matter, we have not identified any of the persons asking questions or making comments, except when that person was the speaker. This would, we felt, preserve the anonymity of the seminar, thereby encouraging a freer and more open discussion at future sessions. For those who are interested, a list of seminar participants is included at the back of the book.

This volume, representing the record of the Columbia University Seminar on Technology and Social Change during its second year of existence, is actually a sequel to the first year's record, published last year by the Columbia University Press under the same title as that of the seminar. Students in this field might well wish to consult that volume, edited by Eli Ginzberg.

The editors note with appreciation the arrangements made by the U.S. Department of Labor—Office of Automation, Manpower, and Training—to disseminate broadly copies of this report, thereby assuring a larger audience for the views and opinions expresed in these discussions.

AARON W. WARNER
DEAN MORSE
ALFRED S. EICHNER

Columbia University
February, 1965

Contents

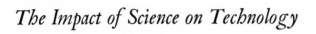

The Impact of Science on Technology

Introduction

by CHRISTOPHER WRIGHT

Executive Director, Columbia University Council for Atomic Age Studies

As MODERN SCIENCE advances, remarkable opportunities emerge for the discovery of still further scientific knowledge and for society to make effective use of such knowledge. Moreover, it appears that some of the principal needs of our society can be met with the aid of advanced science-based technologies. The character and importance of contemporary interactions between science and technology is affected by awareness of these many opportunities and pressing needs. The interactions may be complex and even contradictory, but they would also appear to be increasingly subject to policy choices and deserving of clarification through discussion.

Each of the seven papers forming the core of this volume is addressed to the problem of reconciling opportunity and need in this context, without resort to definitional or other verbal analyses which might seem to dissolve the practical problems. Together they serve to highlight areas of agreement and to clarify the current debate among thoughtful men of affairs who have had specific but different experience mediating the interactions between science and technology. It is, perhaps, significant that in each instance the problem is cast in terms of the attitudes and special capabilities of outstanding individuals and private organizations and of the need for more comprehensive understanding of the problems and prospects. The general burden of proof is shifted from advocates of social innovation to those who tend to assume without question that rich civilian benefits will come as a by-product of sciences and technologies oriented by military considerations; that all geographical regions will benefit from scientific enterprises supported without regard for their geographical location; that industrial, financial, and managerial units operate on a large enough scale to meet the opportunities provided by science; and that the technological facilities of the nation are sufficient to make full and prompt use of discoveries in the basic sciences. There is less agreement, however, on an optimum reconciliation of scientific opportunities and social needs inviting technological solutions or even on whether, as a practical matter, the interactions should be guided by scientific, political, or traditionalistic considerations.

Because the subject is timely and yet insufficiently analyzed, it elicits strong and sincere opinions reflecting personal experience and responsibilities. Some of these opinions and viewpoints are presented here in the belief that their informal juxtaposition will forward analysis of the subject. This was the purpose of the Columbia University Seminar on Technology and Social Change when it invited and discussed the presentations which appear in this volume together with summaries of the related discussions. It is also the purpose of the concluding essay in which the chairman of the seminar, Aaron Warner, identifies various themes warranting further inquiry or systematic research.

Taken together, these presentations serve to emphasize the difficulty of reconciling many of the ideals of science and of social progress. The uniqueness and value of science as a social enterprise is now widely appreciated and yet, as a practical matter, the institutional isolation of science and scientists from local or national needs is rejected even though it is now claimed that this tradition may be essential if science is to make enduring contributions to mankind.

The presentations by I. I. Rabi and Harvey Brooks focus on the interactions of science and technology as seen from the viewpoint most closely associated with scientists speaking for science and the intellectual opportunities which it provides. Two officials of the federal government, John Brademas, a member of the House of Representatives from Indiana, and J. Herbert Hollomon, Assistant Secretary of Commerce for Science and Technology, emphasize the ways in which social and economic needs and political processes must affect the interactions of science and technology. The three presentations by Alvin Weinberg, Ralph Halford, and Frederic de Hoffmann focus on policy matters relevant to the specific institutional problems and solutions of three quite different types of organizations each responding to the changing relations between science and technology. Each reflects his experience as a senior executive accountable for the responses at, respectively, the Oak Ridge National Laboratory, Columbia University, and General Atomic, a division of General Dynamics Corporation.

With the exception of Mr. Brademas, who received advanced academic training in the social sciences before becoming an elected political official, the contributors were all trained as natural scientists. Similarities in their experience and views attributable to this fact do not, however, obscure major differences in outlook and emphasis affecting their judgments as responsible executives or advisers concerning the science affairs policies of government and private organizations. Representative Brademas does, indeed, call attention to the dilemma associated with the scientist's stress on precision and the politician's acceptance of ambiguity. Whether or not this dilemma is ultimately resolvable, the following presentations serve to indicate some trends in thought and practice aimed at satisfying both the demand for more precise statement of the increasingly complex relationships between means and ends involving science and technology and the demand for new kinds of activities, programs, and plans which may indeed seem ambiguous and confused if viewed in purely conventional terms.

Thus, Mr. Rabi and Mr. Brooks both stress the need to recognize and exploit opportunities for advancing the frontiers of scientific knowledge. They share the conviction that such knowledge is an essential basic social resource and may even be the only unquenchable source of ideas leading to new values and goals for a satiated society. And yet, they offer significantly different interpretations of the autonomy of science. Mr. Rabi suggests that science requires unique aptitudes apart from and perhaps even inherently incompatible with the capacity for practical invention or innovation. Mr. Brooks suggests that for the most part the difference between scientific and inventive tendencies is attributable to an alterable difference in attitude and purpose. One view emphasizes basic differences among individuals, institutions, and even nations with some serving as wellsprings of new fundamental understanding of nature and new techniques of inquiry without regard to consequence, while others serve as sources of grand inventions or mobilized innovative capability. The other view suggests that a scientific approach can and should be effectively applied over a wide range of intellectual and practical activities. Since science is not predestined to advance in one particular direction nor to be exploited in any one technological context, Mr. Brooks also suggests the need to clarify and resolve planning and policy disagreements involving scientists with different views in an informed and enlightening debate, which will lead to a pragmatic balance, if not a perfect reconciliation, of scientific opportunities and technological responses to social needs.

A complementary awareness of the need for new forums for debate and decision and new kinds of private institutional initiatives is suggested by Mr. Brademas and Mr. Hollomon, speaking as government officials charged with specific legislative or executive responsibilities at the national level. Mr. Brademas reminds us that our intellectual and material resources must be viewed in terms of political interests, but that this need not imply an emphasis on monetary budgets to the exclusion of manpower budgets. Resources must be developed and distributed with the help of political processes involving vigorous lobbying and bargaining as well as better information and understanding. In fact, the people's representatives are rapidly becoming aware of the local importance of science centers and highly technical industries but this does not necessarily mean these are valued as the seeds producing new resources for the nation or the world. The prospect of general but indirect benefits resulting from advances in science, technology, and related human abilities is less impressive and may be too readily discounted in the political process.

In addressing himself to the problem of making a rich scientific-technical future a matter of direct political concern, Mr. Hollomon stresses the need for specific civilian technologies and supporting basic science inquiries. He reiterates the call for more direct national awareness of the objective of specific civilian well-being, as distinct from *laissez-faire* efforts to meet the objective as a by-product of scientific and technological efforts initiated for other purposes. He suggests a need for relevant systems analysis and planning and the large-scale mobilization of scientific and technological efforts and institutions. In speculating on how this might be achieved, Mr. Hollomon points to the possibility that existing industrial giants might, in concert with the government, use their great resources systematically to "invade" backward sectors of the civilian economy—to build new cities and markets for new kinds of enterprises and products—all on a scale commensurate with the needs of our urbanized society and sufficient to make best subsequent use of the resulting experience with successes and failures.

But the new kinds of information, interests, and organized viewpoints drawn into the political process are not likely to leave this process unaltered. The least one might expect is a shift in the weight which elected politicians and their constituents give to the factors revealed in the course of more comprehensive and systematic analysis of social needs and technological responses. Indeed, Mr. Brademas' own account illustrates the depth of analysis which a perceptive politician can take into account and use in the political lobbies. It also reveals a concern about the unnecessarily narrow views which persons may have imposed on themselves—whether they identify their interests with a particular business, industry, locality, or professional activity.

In the past, a relatively small number of leading American universities provided the principal home for modern scientists. Dean Halford reminds us that industry and government now depend on these universities for necessary intellectual assistance and technical manpower. This view is reinforced by Mr. Weinberg's concern that university science may become too baroque and university scientists too effective in establishing the pursuit of pure science as a superior national goal. It is also reinforced by Mr. de Hoffmann's observation that a key problem for a science-based industrial corporation is to assure its leading scientist-executives that they are, in effect, professors as well as executives alert to the technical opportunities provided by advances in scientific disciplines. The debate implicit in these presentations thus centers on the quality of science, the utilization of scientific manpower, and the problem of priorities within science. Can the universities best perform their unique social function by emphasizing the need for protection from the vicissitudes and distortions of external financial support, the tradition of autonomy, and the ideal of science apart from technology? Can they continue to insist on general support to cultivate self-directed scientific enterprises and training programs and coincidentally perform the instructional and research roles which would be assigned to them in any explicit analysis of their social functions? Must profit-making industries and the government insist on emphasizing immediate needs at the expense of long-range opportunities or can the risks and returns on investments in science and technology be distributed more equitably?

The contributors raise these questions and advance tentative solutions, for while it is clear that scientific research has won well-deserved social recognition, it does not follow that previous relationships can be maintained. Nor are the traditional sponsors and specific beneficiaries of science necessarily prepared to accept the structural changes and other consequences of integrating research activities within industrial and other organizations. Mr. Weinberg and Mr. de Hoffmann both emphasize the inhibitions resulting from the usual institutional and manpower categories and the latter also deplores the tendency to strive for specific and frequently grandiose or glamorous technical or scientific objectives without adequate appreciation of the need for intermediate and adaptable "building blocks" which would constitute new categories of competence and knowledge involving an interplay of disciplined scientific explorations and problem-oriented technical investigations.

From now on we must, of course, expect continued questioning of the efficiency with which science-oriented academic research, and industrial organizations contribute to the definition and fulfillment of specific social goals. But the contributors to this volume and the participants in the seminar also seem to suggest that there are resolvable doubts about the best procedures for improving efficiency. To what extent should the universities and the newer supporters and users of science and scientists be guided by history, by the results of political and economic bargaining, or by the suggestions for coordinated effort emanating from systematic analysis. The Columbia University Seminar on Technology and Social Change does not, of course, answer even such basic questions, but by drawing its members and discussion leaders from a variety of disciplines, professions, and organizations it does endeavor to clarify significant political issues and opportunities for relevant research.

This particular inquiry into the interactions of science and technology occupied the seminar during its second year, and was an outgrowth of the seminar's initial delineation of various sets of basic problems in which its members have an interest.[1]

---

[1] The record of the first year's proceeding was edited by Eli Ginzberg and published under the title *Technology and Social Change* (New York, Columbia University Press, 1964).

The Interaction of Science and Technology

by I. I. RABI
University Professor, Columbia University

I MUST SAY that while I was being introduced, and told of all the committees and things I've worked on, I felt a little guilty. I should be able to give a better organized talk on the subject than I will give. I should do better, considering the nature of my experience, because I spent five years during the war working on the development of microwave radar. I started with a laboratory of about 30 people and finished with 4,500. It was a very effective laboratory, if you count the destruction we wrought on the enemy. I was also closely connected with Dr. Oppenheimer and the development of the atomic bomb. I knew the chief people concerned with the transistor, as well as Charles Townes who developed the Maser at Columbia in the laboratory I set up. I was close to the people, beginning back in the late 1930s, who developed the Klystrons. So, you see, I have been close to some of the major scientific advances in recent times. But as I say, I've not formed any theories on this which I could give you to criticize.

I'm not sure that one can generalize about science and its applications. It may be a more practical problem, from which no general statements can be made, where you can find a hundred counter examples to every example you can put forward in support of any one theory.

It is true that science and technology do not go well together, although they are of the same breed. It is a little bit like
the race horse and the draught horse being hitched together for
practical purposes. In most cases I've seen—except during the
war—scientists who have devoted themselves to inventions fail
to be scientists after a time. The famous Michael Pupin at
Columbia was one of the pioneers in American physics. It was he
who brought back from Europe the "new physics" around the
turn of the century, but he invented the loading coil, which at
that time made long distance telephony possible, and he never
got back to science. I could cite other cases. Nevertheless, the
scientist and the inventor do have to work together to bring
science to practical results.

We live in a complicated age, in the sense that at no time in
history has science flourished so greatly and has interest in science been so profound and so widespread. At no time have the
scientific tools of thought been so basic to all thought. At the
same time our intellect has perhaps been dulled by the successive
applications of science to technology which have so changed the
material world, and even the social scene, in recent times. Scientific knowledge has now become the possession of an elite, and
its diffusion has become less successful. Why this should be so is
an interesting subject in itself.

The process whereby scientific knowledge is translated into
technology involves a succession of quite different kinds of persons with quite different kinds of imagination—the scientist, the
inventor, and then the technologist who can reduce an invention
to some practical form. Then you have the further problem of
production, which takes another kind of engineer, another kind
of imagination.

Sometimes the same person can be both an important scientist and an important inventor, but I don't think there are many
such people. I have a feeling that if you invent, if you continue

to invent, you won't be a scientist for long. The two activities use different parts of the brain, I feel. But we have to make some distinction between scientists, because many different kinds of activity are included under the generic term science.

There are those scientists who are concerned with broad and basic principles. There are the inventors of scientific instruments. The latter are extremely important in the further investigation of various questions which lie very definitely within the realm of science. There is still another type of scientist who not only engages in the careful study of phenomena but also discovers new facts about them. Then there are those who develop techniques. These may be physical or chemical techniques, or they may be problematic techniques which enable one, for example, to go from broad principles to definite cases. Again these may not necessarily be the same people. The great generalizer may not be at all skillful at specific cases. Yet all these different people travel under the same rubric, science. The characteristic common to all these individuals is, in a central sense, their motivation, this being to gather and organize knowledge. In pursuing this sort of activity, they all have an instinct for the jugular vein.

Once you have distinguished between these different types of scientists, the question is, how do you get the result of their scientific effort translated into more practical things? What some of my colleagues and I have found—I am now going back twenty years and I'd be interested in the experience of others nowadays—is that when some very good man discovered a new basic technique, a general broad advance in the art, there was no lack of inventors to utilize it. As soon as a breakthrough was made in the understanding of basic principles, like Maxwell's equations, as soon as this understanding could be translated into more familiar forms that had been used before, such as resistors and capacitors, then all sorts of inventions would come along.

The original man who did the basic work had little connection with this subsequent process. He wasn't terribly interested in it; his problem was a different one.

When the people who are working along in the basic things *do* make something clear, and the people who develop techniques are present and are in communication, we find that inventions simply bubble forth. Our great difficulty was—this will be the experience of many people here who have worked in industry—to keep the inventor from wishing to go on with his own invention and trying to make it practical. We've had enormous difficulties with that. A man who made a brilliant invention wanted to make it practical, but his talents were not in that direction. He had the original concept—it was all there—but to go ahead and make the thing work required a different talent. He would never have the patience to supervise all the details.

To illustrate this point: I was in charge of a laboratory engaged in the development of the magnetron, which was fundamental to the development of microwave and radar devices. We had to make magnetrons by the thousands, which meant the magnetron had to be standardized. But before that could happen, we had to be able to describe it. This may seem like a trivial problem, but it took quite a few men months to find out how to describe it, to reduce it to drawings, so that it could be properly made by the manufacturers.

I might add, after this had been done, I went to Washington and reported to the chairman of the committee that had asked us to do it. I said, "Well, I'm happy to tell you we have it all standardized."

"Oh, that's very nice."

"What shall I do with it?"

"Oh," he said, "it's not my job. You bring it over to the Signal Corps."

I went to the Signal Corps and spoke to one of the officers there. "No," he said, "it isn't my job. You go and see Colonel

X." I brought Colonel X the happy news that the magnetron had been standardized.

"Oh, that's very nice."

"Well, what shall I do with it?"

"Oh, our group works only on nomenclature."

By this time I'd had it. No matter what the authority, my colleagues and I decided that this was what the magnetron was going to be, this was how it was going to be manufactured, and that was how it was done. I don't know whether they have the same sort of problem in private industry, but I suspect they do.

That was not the end of our troubles. We still did not have a magnetron in production. We had only the prototype constructed in our model shop. Our next task was to break down the various stages of production into simple procedures so that the magnetron could be made economically and in such a way that each one would work the way it was supposed to. You can see that in a certain sense, the conception of an invention is marvelous, but as you continue to press forward, it becomes more and more solid routine work. And it's fortunate that human beings are so variously made that there are those who prefer to do the latter.

Here is where the art of organization comes in—to pair off the people who do these very different things, even though each one thinks that he is doing *the* important job, that if only the other fellows would let go and allow him to do his job, all would be well. It's like a symphony orchestra, with all the disparate elements you have. If the various groups are not brought into harmony, the organization doesn't function.

I'm not sure that science has been so terribly important for a lot of the basic technology we have today. Just thinking about it recently, I recalled the older arts and the great progress that was made without science: the wonderful metals, before the knowledge of chemistry or metallurgy was even developed; the wonderful fabrics that were made; building materials; sailing ships, really the most improbable and remarkable things in the world (much more so than steamships); techniques of mining, agriculture, and forestry; the arts of war without the help of RAND and the arts of statecraft without the help of Brookings; the preservation of foods; the utilization of water power, wind power, horsepower—I mean the tremendous advance that was made from the time of the ancients to the Middle Ages in the harnessing of horses and the domestication of animals; dyeing, weaving, pottery—all these things were developed to a very high degree without the help of science.

When we look at the history of industry, it seems to me that science had very little effect on industry, that is, on the older arts, until it developed to a very sophisticated point. The steel industry, for example, made very little use of science until recent times. The art of making steel developed by itself without the aid of science. There might have been some elementary data, some primitive thing of that sort, but no more than that. On the other hand, when you turn to other industries, like the electrical or some of the chemical industries, many of which developed out of basic scientific work, then you have industries in whose development science has been essential. The prime effect of science has been, it seems to me, to beef up some of the older arts, but really, more importantly, to make new industries, add new knowledge along with new components and new techniques.

A very important element in the growth of industry and technology is the adoption of instruments initially developed for scientific advance itself. I was really astonished the first time I saw a mass spectrometer being sold commercially. It almost seemed sacrilegious—to think of a real laboratory device being offered for sale to the general public. But such devices have found enormous application in industry, specifically nowadays in atomic energy, with the use of isotopes and other devices initially invented for scientific purposes.

In recent years, particularly in the postwar years, a whole new class of tremendously powerful scientific devices have been invented, practically none of which have found any application; yet I feel that somehow or other they are so basic and so important that they may determine whole important technologies of the future that people at present cannot foresee. I mean, for example, the devices which have been invented for the detection of the phenomena in particle physics, things like the alternating gradient machines, the big accelerators—machines encompassing new principles of focusing particles of very high energy which have changed the whole outlook in that field of scientific investigation. Formerly if one wished to double the energy one needed eight times the velocity, but this new principle has changed the relationship completely. It is now linear. The machine at Brookhaven, capable of producing 33 million electron volts of energy, would have taken the whole national treasury to build before the development of this new principle.

The transistor is another example. Until what I call the detailed science of material properties had advanced to the point where materials could be purified, the modern transistor couldn't be made.

There are scintillation detectors, there are cloud chambers and bubble chambers, there are high-frequency circuits with which one can comfortably make measurements of events which take place in one trillionth of a second. There is something most of us have never heard of, called Cerenkov radiation, after a Russian who first observed the phenomena.

There are other things which are now very esoteric but which are bound to enter into industry. It is merely a question of improving technology to make the transition from the basic scientific observation to technology. One such phenomenon is superconductivity in materials. One can see over the horizon electrical apparatuses now the size of a house becoming quite small and handy, perhaps with the same capacity as the internal combustion engine. I can foresee a tremendous advance in basic science and in the instruments of science, resulting in methods and techniques on which important new industries will be founded.

Interestingly enough, the industrial companies rarely if ever develop these new devices. They are developed almost entirely by those who work in universities, or now, in special laboratories such as Brookhaven. Then they seep into industry in indirect ways.

In this area of developing new scientific instruments, there is a brand new phenomena in the United States. That is the practice of scientists and engineers getting together and forming small companies for specific purposes. There are any number of these specialized firms. Various associates of one of those companies, for example, produced the Klystron.

In the United States we now have a large number of people who design and make instruments in a practical kind of way. This is a tremendous advance, because when I was a young student at Columbia, very few instruments were made in this country. If one wanted a good magnet, one bought it abroad. If one wanted fine optical instruments, one bought them abroad. In fact, any instruments that were scientifically sophisticated had to be purchased overseas.

Now the situation has turned around. The laboratories in Europe, India, or elsewhere, are stocked with American-made instruments, which are the by-products of the scientific work done in American laboratories.

We are in the midst of great flux. The ability to make these new instruments, mechanical, electrical, chemical, and others, is becoming more and more widespread. We have the institutional setting to go from scientific discovery to practical application, not only within large industrial concerns, but also through these other small firms.

What we do not have to the same degree as other countries

are inventors in the grand style, those who translate science into technology. And I have a feeling that somehow our particular culture does not encourage originality and eccentricity, which are often the mark of the inventor. These qualities do not find their place here to the degree, for example, that they do in France, or even in England. Our system of education is superb, but it is a system geared for large numbers. Not that the maverick is not tolerated; it is just that he does not have the support of the community. He keeps alive, but he feels uncomfortable all the time.

It reminds me a little bit of a friend of mine many years ago who went to the University of Minnesota to teach theoretical physics. He wrote me, "I can't really do theory out here, because I feel I ought to be out shooting timber wolves."

What I'm talking about are inventors who make large breakthroughs in the arts, the actual honest-to-goodness inventions. Consider, for example, the jet engine. Although we make the best in the world, it was not invented in the United States. Or take microwave radar. Our laboratories really made a significant contribution to the development of microwave radar, but the principal elements of the system were invented elsewhere. Both of these inventions constituted large steps in the advancement of technology. They are the sort of thing that I am talking about.

Once we get over the hump of the grand invention and begin to get into its development, we can do a better job than the peoples of just about any other nation. Actually, I think in most instances, if you look back far enough, you'll find some Frenchman who is behind the grand invention. I don't think it's an accident. I think it reflects the quality of French education.

One advantage which we have in the United States, in this case a legacy from World War II, is a love of teamwork. I believe that half a dozen people of good quality working together are the equivalent of one person of very superior talent.

So that in comparing ourselves with the British, I think it's true that one Britisher is worth three Americans. But Americans working together can run away from them.

I'm not sure that, in the technological area, talent by itself is the key to competing in the long run. For example, the French fundamentally are more original, but they're also more individualistic. They don't work well in teams.

I'm not sure that technology should proceed as fast as it does. I'm not sure that it's doing much good, in the sense that I don't think we understand all the social implications. I'm not sure it isn't doing things to us that we may later recognize as undesirable, in terms of our own standards at the present time, whether it isn't destroying values which we feel very deeply about.

Finally, let me say that I think that the development of science is central for the continued existence of human society. Suppose we extrapolate from where we are now. By proper organization, which is not too difficult to achieve in the United States, people will, in the not too distant future, be fairly well assured of a reasonably comfortable existence.

What, in this future, do you have that will excite a man to give his life some meaning? It can't be just another motor car. People can't eat more than a certain amount. They're already consuming too many fatty foods as it is. What will excite young people? What will their imaginations aspire to? What frontier is there, if it's not the frontier of understanding the environment, understanding their own nature, understanding the nature of matter?

That is the great thing for us to do, and if we don't do it, life will become humdrum, and people will rebel out of boredom. So it seems to me that the future of society—if it is to have real life to it—is in science, and I'm not talking only about physical science. There's a whole range of problems that call for a rational way of investigation. I don't see how we can neglect this without reducing the quality of civilization in our country.

Unfortunately, in some of the areas outside the physical sciences, progress has been slow, but essentially one doesn't advance with equal rapidity in all areas, and it's not even certain that more effort won't at times put a person back. Very often the advancement in a field depends upon somebody getting *the* important idea which is the key to it. Professor Butterfield at Cambridge, in a book called *The Origins of Modern Science,* points out something which was new to me, that the ancients thought that the natural motion of a body was circular. It wasn't a stupid idea. This is what they saw happening to planets, and so they saw it as the natural motion. But to start with that, and then try to explain straight line motion, is impossible, and it therefore took two thousand years to overcome that idea. Many of these questions sometimes must wait a whole generation, or several generations, to find the key idea, the one that leads to a breakthrough. The ancients, with all their brilliance, were unable to develop science. The Chinese, with their large numbers, great abilities, and enormous skills, having reached a certain point, were unable to advance further. How in the world did the barbarians of Europe, the kind of people who were civilized late in human history, develop this art and this technique?

As for the arts and the imaginative understanding of our world which they give us, I prefer not to generalize about fields in which I have no competence, but I do consider them in some ways quite different from science—not contradictory, but different. I see no reason why anybody who has talent for artistic activity and wants to engage in it shouldn't, if it gives him satisfaction. But the understanding which it gives is not the same understanding that science gives. It doesn't have the social influence. It doesn't have the possibility for commonality of understanding that science has. The arts depend much more on local conditions. I don't think these kinds of imaginative activity are at all contradictory. And I think it will be easier, as time goes on, for men to cultivate both kinds. I mean, as we increase our general prosperity and well being, there will be no difficulty in devoting ourselves to both science and art.

The arts are certainly central to life. Yet they are not the kind of thing that will inspire men to push on to new heights. Suppose we were to become a nation of poets and were taught in school, as the Japanese are taught, that every good citizen should write a poem. Some would be very good and people would read and enjoy them. But what would anybody talk about? Only everyday things—love, sorrow, life, and death. If men want to go beyond these everday things to a grand theme, they will find it only in science.

Even the works of Shakespeare, which are essentially an exploration of human character, are really wonderful, glorified gossip. I am not degrading gossip, because we live by it, but it does not take us outside ourselves, outside the human race. Art works in a different realm of originality. I mean, after Shakespeare and others, how many more persons can say the same things? They can say it in different ways, they can say it in beautiful ways and in other contexts, and I hope they will never stop. But just the same, where do we find the really new thing, the moving thing, the thing that will show the glory of God and the originality of nature, the profundity—and where do we find the imagination that delves into the mysteries of life, into the existence of matter? That imagination, I would say, will be found only in science.

DISCUSSION

THE EXTENT TO WHICH SCIENCE CAN BE DIRECTED

Question: The British Labour Party has recently announced that science is to be the foundation of socialism. It hopes to link science and technology in an effort to increase the nation's productivity. To what extent can you direct scientific research, particularly to channel it more directly into technology?

Mr. Rabi: I think that to some extent it can be directed. But it's very difficult, and perhaps even a contradiction in terms. The Russians have tried it in varying degrees. They've tried to direct research, fundamentally, through their academy with all its branches. What I've seen of it has not been very good, especially when the quality of the people involved and the magnitude of the effort is taken into account.

Soviet science is too hierarchical. The Russians have tried something similar in the arts, and I think the success has been of about the same kind. They've got some people who can write poetry, novels, and motion pictures along the lines of social usefulness, and I think mostly they disappoint. It's not that they don't have the people, but that science, like the arts, does not lend itself to direction. It can be furthered through motivation to some degree but the situation in Russia, it seemes to me, is like that of all planned societies. In principle, they're marvelous, but the director doesn't have enough information to understand the situation in detail.

I doubt if we will find—as scientific efforts increase—a corresponding increase in people of high talent. I asked the Russians what happened in their case, what the rate of increase was, and they said that it was very, very low. It seems that if you multiply the number of scientists by ten, you will wind up with only twice as many bright people.

Question: I would agree with the broad sketch. I think most of us would. But the growth of so much scientific research involves huge outlays. For example, alternate gradient machines cost $20 million each. There's a need to establish priorities somewhere. My question, then, is who is going to set these priorities and on what basis are they going to make their decision?

Mr. Rabi: It's a question which is bothering everyone. In the first place, we have to get accustomed to scale. If we realize that science and its applications are more or less central for the further development of the economy, not to mention maintenance of our national position, these sums appear trivial. After all, let us say a machine costs $30 million to build and install, and then $5 or $6 million a year thereafter to operate. It's nothing for the United States to have five or ten such machines. These sums seem large only to those who want to get something for nothing or for very little. There are always attempts to drive too hard a bargain with these things. Everybody talks about the great blessings of science, but when they're asked to put their money on the barrel head, they say: "I'm not sure. It costs too much." Such an attitude permeates the whole society.

The assessment you ask for is hard to make. I don't think it can be done precisely, of course, but there is a sense within the scientific community of what is important and what is not. This judgment won't be correct all the time, but since this is a human adventure, I don't know how we can do better.

Question: Aren't there some scientific areas which, if greater resources were devoted to them, would generate more noise than information?

Mr. Rabi: I think that if the right man were in charge, he would know which areas those are. It's no secret within the scientific community, though it may be a secret to the man who is handing out the funds.

Question: Would you say something further about the role motivation plays in the direction of science?

Mr. Rabi: I think that is a matter of tradition and of fashion. I wish some sociologist would explain fashion to me, whether in intellectual ideas or in women's clothes. It's a phenomenon which dominates human behavior, and very few of us ever escape from it.

There are fashion setters in science just as there are fashion setters in women's clothes. They set the style in intellectual endeavor, no matter what the scientific field. Fashion can be swayed to some degree, for example, by the nature of certain rewards. At very small expense, the members of the Swedish Royal Academy determine in part the directions of scientific effort through the Nobel Prizes which they award.

The direction of scientific effort has to come in some way through the natural leaders of the community, and that can be done. I will give you an example from my own experience. I left Columbia right after Election Day, 1940. I voted for Roosevelt and then went to Cambridge, Massachusetts, to be one of the founders of the laboratory there. It was a year before we entered the war, but we saw it coming.

About thirty of us started working on microwave radar, but then we needed more people. So I went the rounds to different universities throughout the country. When I'd see some young man I wanted I'd say to him, "Now, look, we've organized a little laboratory in Cambridge. The living conditions are not very good, but I think it's important. I can't tell you what it's about, but I'd like to see you there in two weeks." And I wasn't turned down once! This was no trivial thing, for in the end we shut down almost all physics research everywhere else in the United States for the rest of the war years.

Now, why did those people come? I didn't give them a long song and dance. I remember one fellow I talked to who was very anti-Roosevelt. He said, "That man is going to get us into the war. I'll bet you he's going to get us into the war."

I said, "Let's not argue about it. You just come on." And he came. When we got into war he said, "I told you that man would get us into the war."

Why did he come? Of course, there were many reasons, some of them not applicable in all cases. For one thing, one could see the direction which events were taking. For another, I was one of the Grand Old Men of physics. I was only forty years old but well known. We had, in that sense, when I went the rounds recruiting personnel, a natural leader of the group. We'd more or less come from the same bar. We knew each other.

We made the rule in the beginning that salary would have no relation to the position a man occupied in the laboratory. In other words, he might be the boss of the group even though he was a young man who'd been only an instructor previously and $2,500 a year was enough to hire him. And under him might be another man who was much older and who had been a full professor at a much higher salary. We could make that stick for a while—though not terribly long. But it was all done, in a certain sense, in the way the British keep discipline.

I do feel that somehow or other, with proper sensitivity, the direction of scientific activity can be organized. As a matter of fact, Emanuel Piore did a marvelous job in that field through the Office of Naval Research without interfering with anybody. In a friendly, almost humble, kind of way, he was able to direct efforts so that we were ready for certain situations which he had foreseen long before those situations finally developed. But it requires that sort of ability—to foresee events.

When substantial effort is poured into a field, interesting things often do come out. And then people will go into it, naturally, because it has an interest of its own. Things get started in that way.

When I was a young man, solid-state physics was far from an interesting field. One man, after working in it for twenty years, quit, saying, "Well, there are no general properties." Now, of course, it's one of the most exciting fields of scientific research. That is the way it happens. I do not believe that real organized effort, with the kind of people who generally get to the top of such organizations, would do very well. I do not believe that the creative end of life can be fully organized— partially organized perhaps, but not fully.

Question: Isn't it true that the government, by deciding to embark upon a particular project, may determine the direction which technology and science will take, with possible undesirable consequences. I'm referring to the current space effort, especially the efforts to reach the moon, and wondering whether this doesn't represent a distortion of national objectives.

Mr. Rabi: In that particular case, I doubt it. Scientifically the moon is not a very interesting subject. It's nowhere near the center of science. I think it comes more under the heading of exploration.

The fact is that the space program does not get the best people. It's just something which is going on in the basement. It's using a lot of material. It provides headlines. But actually, it has only peripheral scientific interest, except for astronomers.

Question: Could you say that space is the moral equivalent to war?

Mr. Rabi: That would be great. But more concretely, many people have felt that from an economic point of view we will need alternative areas if there is to be a radical reduction in government expenditures on armaments. First let's move into the area of space exploration; then we can move into other areas.

Comment: Going back to your description of science in the Soviet Union, I get the impression that while it has had certain successes, these successes are limited and will continue to be limited by the way the Russians have organized and structured their scientific establishment.

Mr. Rabi: I compared their successes to their effort; I was not trying to belittle their achievements. Actually, I don't know too much about the Soviet Union, since I've been there only very briefly. But I could see the results. Moreover, I met a number of qualified Russians, and I talked on just this question to some of their leaders. And I am simply reporting on what I observed. I had the feeling that Soviet science is too hierarchical, that the director actually directs too much. This is not just peculiar to the Soviet Union. It's European. Most of the things I would criticize in the Soviet Union are also true of Europeans, and Europeans in American universities.

In the Soviet Union, there's a certain type of provincialism, which I can readily recognize because it is similar to what I knew in the United States at one time. The Russians I met were people who knew the literature thoroughly, but they were not really in the mainstream of science. Many Americans have noticed, when they go over to Russia, that their Russian counterparts know their own work, the Americans' work, better than the Americans themselves do. The Russians are thoroughly *au courant* with the literature, they know what happens, but they don't feel entirely free to venture on their own. I don't know quite how to describe it, but I think anyone who has been to the Soviet Union has observed the phenomenon.

There is in Russia a hierarchical organization, in which it's difficult for a young man to strike out on his own. Our democracy, as I've seen it, is really marvelous. At most of our universities, the good ones that I know, the rank of a man, whether instructor or professor, hardly matters. Even in my time, when I was an assistant professor, I had more students than any of the professors. There was no difficulty about it then, and there's even less now. That would not happen in Europe and that would not happen in Russia.

Question: Do the Russians themselves feel that way about it—that they're too hierarchical?

Mr. Rabi: I can't answer that. I'm just describing the situation from the outside. I could see how thoroughly everything was organized. One of the cyclotrons was being used most efficiently. The floor was covered with different kinds of experiments and everybody had his allotted time. But I don't know how a man with a different idea would gain access to one of the machines.

Comment: In Europe, there is this extremely hierarchical organization, but the young ones are now rebelling.

Mr. Rabi: Yes, they're rebelling, and I think the young people in Russia will, too, especially if they're allowed to travel to other countries and see a different way of life. Then they might go back and try to do something about it. I think Russians really suffer from their closed door policy.

Question: Yet they have a high degree of success. How do you account for that?

Mr. Rabi: It's a tremendous national effort. I only said, relative to the effort—and it is a tremendous effort—the results were not strikingly good.

THE INFLUENCE OF CULTURE

Question: Dr. Rabi, you said that we have a short supply of inventors in the grand style. But we seem to have no lack of inventors that can translate ideas into practical results. What kind of institutional arrangements are responsible for giving us this adequate supply?

Mr. Rabi: I think we've made one of the greatest of all social inventions, the existence of widespread higher education. The graduates of these institutions may not be tops, but a large number of them at least reach a certain point. Then, we have the graduate school, which is something new under the sun. It takes a fraction of that large number and brings them forward in a systematic way. The difference between the Europeans and us is that the Europeans have the cult of genius. The exceptional man is highly prized, almost worshiped. He is a law unto himself, in his own bailiwick. This is exactly where our democratic attitude has given us a different kind of strength. But it has been, to some extent, at the cost of the other. What we need to do is to try to find a way to get both qualities.

Question: I was wondering whether there isn't an additional factor in the Europeans' cultivation of the genius? I agree that you cannot organize geniuses. They have to be left alone. But I think there is an additional feature which is very important—

the existence of technicians who know enough about the basic sciences to perhaps understand the genius and translate his discoveries into new technologies. In Europe the technicians do have this type of knowledge, while in the United States the technician is a technician, pure and simple. Most of the time he is unable to understand the contribution of the genius. Couldn't we in the United States, perhaps in our universities, raise the level of our technicians to the point where a few of them might become the interpreters of pure science to the other technicians?

Mr. Rabi: Just in my own lifetime, the United States has changed so much in this respect that it's almost unbelievable. In 1927, when I went to Göttingen, Germany, the university subscribed to our journal, the *Physical Review*, but they waited until the end of the year to get all the issues, in order to save postage. Ten years later, it was one of the leading physics journals in the world, and now it is by far the leading physics journal in the world. I doubt if the university still waits till the end of the year to get all the issues.

In my time, 1927, it was not really possible to get a first-rate graduate education in physics in the United States. The interesting thing was that when I got to Germany, I found I knew more, much more, than most German graduate students at the same level. What I lacked was a certain feeling for the tradition. As I have described it on other occasions, I knew the words but not the music, because I had not yet come in contact with the kind of people—a lot of it is oral tradition—who were really making science. My generation brought that sort of thing back to the United States.

Now, Europeans come here to study. It's a matter of great prestige for a German, Frenchman, or Italian to say that he's been to the United States. This whole reversal has come about in a relatively short space of time. The same thing can be done in the case of technicians; we're already beginning to move ahead in this area. I think the Europeans will acknowledge that there's nothing in Europe that's equivalent, for example, to the Massachusetts Institute of Technology. In fact, they've been talking about setting up something like that in Europe, through the auspices of NATO. It would certainly take the combined efforts of all West Europeans to produce an equivalent institution.

Comment: It seems to me that the metaphors which you used in your talk, Dr. Rabi, did not quite jibe with what I inferred from some of your other remarks, that creative activity cannot be organized. You have suggested that genius should be given free reign, unhampered by social pressures. Yet you use the metaphor of the race horses, who must be guided in some measure. I'm not sure but that there isn't some sort of social pressure, some sort of social structuring, behind even the leaders, the creative and scientific geniuses.

Mr. Rabi: Oh, I think there is. I just don't know what it is. That goes for all fashions of this sort.

Comment: I'd like to argue that the genius is the man who writes quartets although the pressure is for symphonies. That's the sign of a genius—he goes against the tide.

Mr. Rabi: That may be so, but you only know it afterwards. If you could identify Mr. X and say, "You're the genius, I won't bother you," or "You received all A's, I won't bother you," it would be fine, but you know that isn't possible. It becomes a circular argument.

Question: Do you think that there is any way the French tradition of originality can be encouraged in this country?

Mr. Rabi: Only at a cost. For example, we in the United States love to work together, teamwork and so on. But there was a time in France—I don't know whether it is still true now—when the student would lock his door for fear that the professor would steal his results.

Question: You kept mentioning the French. Is there anything in the early German development—I mean, up through the Kaiser's period—which indicates a better linkage between academic science at its high development and selected industries like the chemical industry?

Mr. Rabi: Yes, there were trained chemists who couldn't get jobs, and therefore started industries. I think the Germans in the early and mid-nineteenth century had the only good universities. And they produced people who in turn went out to earn a living.

GOVERNMENT AND THE SCIENTIST

Question: You have sketched a very fascinating typology of the modes of scientific imagination to warn against a uniform image of science. You talk about the theoretician, the experimentalist, the man involved in technique. Would you say there is any principle which determines which of these scientific imaginations is best fitted to give advice to government?

Mr. Rabi: I think it's a very serious problem, and we have great difficulty in recruiting people for this purpose. I divide scientists working for the government into two categories. One is a consultant, a person who contributes his expertise to a particular job. He is easily identified. The other is involved in questions of policy. He is not necessarily an expert in any particular area, but he has a feeling for the direction of trends and the impact of other qualities. This second type is exceedingly rare. In the war we did find them, because it was necessary to do so. As we all worked together with the military and the politicians on various problems, those who were capable of offering advice on broad policy gradually emerged.

Question: What was involved in doing a successful job, and developing these scientific advisers, during the war?

Mr. Rabi: During the war we found that we simply could not accept the military's advice on weapons development as the final word, because the military really didn't know much about weapons development. They'd never worked with radar.

They'd never fought a modern war. There were certain things the military could do excellently. They could organize a logistical supply. They could organize men. They had the habit of leadership. They could bring men from here to there. But the new weapons and how to use them, this they had to be taught. We in the United States perhaps had a little advantage, because we could learn from the British who had been in the war before us.

In a certain sense, before we could develop an instrument of war, we had to be able to see several years ahead. Often it would be two years before the new weapon could be put into the hands of the soldiers who would use it. So if the weapon were going to be at all useful, we had to be able to predict the direction the war was going to take. Otherwise we would be working on something that would be quite outmoded by the time it was finished. During the war some of the scientific people developed this knack of sensing out the future trend of events, and those were the ones found in that interlocking directorate, the scientific community in Washington, helping to formulate policy.

In peace time it isn't as easy to spot these individuals, because they aren't working together, all in one place, the way they were during the war. Of course, we now have the possibility of simply waiting to see which scientists are successful in their approach to long-run problems.

Question: In your experience, is the ability to offer scientific advice a knack which is enduring, or do people who have it at one point tend to ride their pet ideas from then on?

Mr. Rabi: The persons I'm talking about can be wrong in their judgments. That's not the essential point. It's their approach to a problem, their ability to take into account many somewhat tenuous factors. And there are others, extremely brilliant people, quite impatient with that sort of thing. They want to see something tangible. They want to get down to specifics, instead of dealing with these vague values and hunches which you sometimes have to do if you're going to help a man who has to make policy.

In other words, you can't just give the President of the United States a lecture on inertial guidance and expect him to decide what to do. It has to be translated into a digestible form. You have to partake of his own thinking. This is political counseling.

SCIENCE AND MODERN WARFARE

Question: I notice running through many of your remarks a certain thread—the importance of war as one of the great forces leading to the utilization of science in technology. I wonder whether there are any social emergencies short of war that have, in your experience, produced dynamic short-term translations of science into technology?

Mr. Rabi: As I said, what has happened in the United States is something which came about very suddenly, just in my lifetime. And of course the war was an important factor. But the period since the war has also seen a tremendous expansion of science. I talk about the war only because it is the basis of my experience.

And certainly the war did wonderful things in some respects. For instance, we began to have money for research. Before the war, the physics department at Columbia was able to spend only $15,000 a year on research. We now spend over $3 million a year. Before the war, a man had to have an extraordinary physical stamina and character to struggle through until he received his doctor's degree in physics at Columbia, for often he had very little money on which to live. We now can afford to support our students adequately. These new funds have enabled a whole new group to enter the sciences.

We've been living, I think, in a very confused period the last twenty years, dominated, in a certain sense, by military expenditures. One has only to look at the budgets of the universities to realize how much of it comes from the government. What makes the government supply these funds? I don't know the answer, but the expenditures have been more or less under the heading of national defense. I think this reflects special circumstances. Certainly we don't see similar expenditures in other countries. As a matter of fact, a good deal of the best research abroad, in France, England, the Scandinavian countries, and Israel, is supported either by the U.S. Navy or U.S. Air Force. I'm talking about pure research.

Of course, I do not know whether the American people, if absolute peace were to descend upon them, would support through their Congress the physical, medical, or social sciences to the same degree they do now.

Comment: It's interesting that the physicists have learned to modify the human gene in various ways, just as the biologists have made a significant breakthrough in genetics. There now seems to be a race to see whether the physical scientists can destroy the gene before the bioligical scientists can learn to synthesize it.

Mr. Rabi: These scientific instrumentalities that I have described interact. The use of X rays and the electron microscope, for example, enables scientists to put together the basic building blocks of the gene and understand how it works. This development affects biology purely in an instrumental way. What the scientist is engaged in is not destruction but understanding. Science has nothing to do with what the practical fellows do with it.

CONSIDERATIONS OF SCIENTIFIC AND
EDUCATIONAL INSTITUTIONS

Comment: To my knowledge, there's never been a scientific research laboratory which shut down when it finished its mission. In view of the fact that the old organizations continue to grow even though their original tasks have been completed, I wonder whether the proliferation of organizations engaged in scientific research is a good idea.

Mr. Rabi: I would like to suggest that every once in a while it might be a good idea to have something like a 30 percent cut—not a small one, because then everybody just hedges a little bit, but a real cut that will force people to make some hard decisions, perhaps even to start again. I do think many organizations have become moribund.

Question: What do you think would happen if the government were to reduce its support of science? Might this not lead to more creative thinking in the sciences?

Mr. Rabi: I think that most human experiences are irreversible. We will never get the 1938 dollar back. We will never be able to go back to the haphazard support of science that existed before the war with practically no government participation. I think the expectations of people have gone beyond that point. So there will continue to be with us in the future many of the large scientific establishments.

Question: Our universities are often, most of them, still structured very much in terms of different departments— business, chemistry, physics, and so forth. Yet most applications require knowledge that transcends these narrow boundaries, requiring many different types of skills. Do you see a possibility of making a better match, shall we say, between our university structures and the applications of science?

Mr. Rabi: I think there is. I think we will have to change our university structures, or at least modify them. I've thought so now for some years. Our universities need a new division which, for want of a better name, I'll call "School of Advanced Study." It should be concerned with interdepartmental matters, a place where new disciplines can start and find a home. Take the history of science for example. It doesn't belong to science, it doesn't belong to history departments. Historians don't know science, scientists don't know history. But if there were a

"School of Advanced Study," such a field would have a proper home—and then it could develop into a self-sustaining discipline.

We have no place where a man who has received training in one discipline can go respectably and learn another, where he can be infected with other modes of thought. Some universities do have a kind of roving professor, affiliated with no particular department, but who ever heard of a roving instructor? We have no place, for example, for a man who has studied deeply, received a doctor's degree in chemistry, and doesn't want to continue in research—doesn't have a knack for it—but would like to study further, whether it be in economics, sociology, or some other field, and then at a later date possibly go into government work, say the State Department, or be assistant to the president of a private company. We have no place in which to train such individuals. In addition to that, we have no one looking at the university as a whole, saying to the various departments: "Are you doing your job? Are you presenting your subject in the context of a certain time and space? Are you humanizing it so that your graduates will leave the university as a part of the total society, rather than as specialists in a narrow field?"

I think such things can be done. I think they will be done, because we're trying to do it now. These seminars, here at Columbia, are part of the method. And then there are the institutes which are springing up all over the country at the universities, bringing people of different disciplines together. But we still haven't found a mechanism to do this in a regular, organized way.

TECHNOLOGY AND ATOMIC ENERGY

Question: Would you comment on recent statements to the effect that very little of the atomic energy effort has been devoted to the development of peaceful applications?

Mr. Rabi: What happened in the case of atomic power is that the very things one could not have foreseen did occur—for example, the discovery of new oilfields that kept the price of oil down. As another example, Europe, after deciding that it had no future unless it could develop atomic energy, formed Euratom. Then, six months after Euratom was formed, new techniques in coal mining created a surplus of that fuel.

But the steps we are going through to develop atomic power we would have had to go through anyway. After all, we can't hope to come up all at once with the sophisticated techniques that are necessary if we are going to meet the competition from conventional fuels, especially with the more efficient power stations that are now being developed.

There's no question but that the peaceful uses of atomic energy are still not realized. But nothing is more certain than that the future central station power will depend on atomic energy. Just when, we don't know, but I am sure it will be before the end of this decade. This will be true even in the United States, unless many new sources of conventional fuels are discovered. But if one stops to consider that the underdeveloped countries may soon begin to develop their own industries and that this is bound to create a real competition for power sources in the future, then one can realize why, by the end of the century if present projections have any meaning, there will really be atomic power.

The Interaction of Science and Technology: Another View

by HARVEY BROOKS

Dean of Engineering and Applied Physics, Harvard University

THE SUBJECT of my talk is really the same as Mr. Rabi's. I feel somewhat embarrassed, following on the heels of what must have been a very stimulating and provocative address, but perhaps each person can talk about the same subject with a little different perspective, with a little different background of experience, and with a little different set of prejudices.

The connection between science and technology, which is really the subject we're talking about, is by and large a recent phenomenon. The technology of the nineteenth century, the technology on which most of our industrial civilization depends, or at least still depended on thirty years ago, grew up relatively independently of science—not completely independently, but relatively so. When one thinks of the technological innovations which have had the greatest impact on our society, one realizes that most of them were products of mechanical ingenuity, rather than applications of science *per se*. One thinks of such things as barbed wire, which was responsible for the opening up of the West; the typewriter, the sewing machine, and the cotton picker; much of the art of metallurgy, building materials, and tools; even a good bit of medicine and surgery.

With the growth of the German chemical industry in the 1880s, a technology began to emerge which was the result of the greater and greater coincidence between science and technology. The four industries in which this has been most evident are the chemical industry, the communications industry, the power industry, and, in some measure, the aeronautical industry. But it is only in the most modern versions of electronics, aeronautics, and applied chemistry that one would say that basic scientific knowledge has really overtaken engineering practice. For the most part, the art has tended to run ahead of the science.

The second point I would like to make is the multiplicity of connections between science and its applications. When one thinks of the relationship between basic science and applied science, one is inclined to think of dramatic models, like the atomic bomb, the transistor, or, to an extent, radar.

The analogy which comes to mind is the analogy of the seed and the plant, basic science being the seed and technology the plant. This, however, is a misleading model; and in fact, the cases which I cited could, I think, be considered exceptional rather than average. A better analogy, I am inclined to feel, is the analogy between the seed and the fertile field. The role of science in the development of technology is to provide the environment in which technological ideas can be exploited, rather than in fact being itself the origin of technological ideas.

One can go back in history and cite many cases in which technological ideas were suggested or invented, so to speak, yet died at birth simply because the sophistication of scientific knowledge was not sufficient to provide a fertile field for improvement and exploitation of the original idea. One finds, as early as 1910 or 1920, many suggestions regarding the use of semiconductors, or what are now known as semiconductors, foreshadowing many of the developments in the 1950s, such as the transistor, the rectifier, and so on. One can see that none of these ideas came to fruition simply because there was not yet a sufficiently sophisticated science of materials and solid-state physics to permit the rapid and successful economic exploitation of these ideas.

An advanced state of basic science may provide the environment in which the new technology can grow, but the basic science itself usually does not produce a new technology. If one looks more carefully at any technological development of this sort, one will usually find that many different strains of science were involved in the development of the technology, not just a single field.

This is, in fact, true again in the case of the transistor, although it's part of the conventional wisdom to regard the transistor as the product of solid-state physics. If one looks at what actually happened in the development of the transistor, one will see that the idea came from solid-state physics, but that all kinds of chemical, metallurgical, and other techniques which had grown up over a long period of time, in some cases going back to the 1900s, were involved in the actual development and successful exploitation of that particular device.

A third point that I would like to make relates to an aspect of the problem that Dr. Rabi emphasized, and that is the relationship between invention and science. Dr. Rabi stated that invention and scientific discovery make use of different parts of the brain, so to speak; that the inventor is a different breed of animal from the scientist.

In a certain sense, this is true, but I think in another sense it is not. I think I would agree with Dr. Rabi if, for the brain, he would substitute the emotions. Actually, I believe that the same qualities of mind go into science, especially experimental science, as go into invention, but in fact it's the emotional attitude towards the subject, towards the activity, that determines whether a man will be an inventor or a scientist. If one looks at the activities of American scientists during World War II, one will see that where sufficient motivation existed, scientists became inventors. What a scientist does when he's creating equipment, apparatus, or anything of that sort for experimentation is indistinguishable in my mind from what an engineer does when he designs a product for the market. I believe that the same qualities that go into invention also go into the creation of scientific techniques. But emotionally these two activities are poles apart.

Now one may argue that this statement is a semantic one, that the net effect is the same, that scientists are scientists and inventors are inventors, but I think that this statement does have important implications in the long run for the relationship between science and its applications and for the way science itself develops. I'll try to elaborate on this point as I go along.

Dr. Rabi has said that historically it appeared, at least in the United States, that whenever an important scientific discovery was made, inventors by the scores appeared to exploit it. This is certainly true with the development of electrical science in the late nineteenth century. It was certainly true with the development of thermodynamics early in the nineteenth century. Nevertheless, I think this is an excessively comforting statement, and I'm not sure that the relationship between science and technological innovation is, in fact, as direct as that statement would tend to indicate.

Science and technological innovation are different things, and one can exist to some extent without the other even though they are symbiotic. There are obvious cases—England, I think, is one—in which a country that is preeminent in science can be backward in technology. Similarly a country which is preeminent in technology can be backward in science, so I don't think it's safe to assume that merely by the creation of new science, technology will automatically follow. On the other hand, of course, it is perfectly clear that in the long run and to an increasing degree in a worldwide sense, the creation of new science is a necessary, though not sufficient, condition for the creation of new technology.

Let me now outline some of the elements or factors which I consider to be conducive to technological innovation originating in science. The first and perhaps most important one is what I would call wide-spread scientific literacy. What one has to have is a population that has a certain minimum familiarity with technical things, a large number of people with sufficient scientific sophistication, if you will, to be successful inventors. If you go back, I think, to the early part of this century, you will recognize the very important role played by the Model T Ford, farm machinery, and the homemade radio in creating a large population of not too well-educated people in the formal sense but a population nevertheless at home with the technology that was being created. I think this was a very important factor in the technological development of this country.

A second factor, of course, not unrelated to the one I just mentioned, is the creation and growth of what I would call a scientific power elite, that is, a scientifically literate management with an instinct for possibilities in the technical area. Again, a comparison of the United Kingdom and the United States, I think, shows a big difference in the degree to which the power structure, to use political science terminology, is infiltrated with people who have a technical background.

The third factor might be stated by saying that to some degree specialization is the enemy of technological innovation. I realize that this is a very controversial statement. Nevertheless, I think it does have an element of truth in it, that the people who create technological innovations are usually people who have jumped fields considerably. The point I'm really trying to make is that I think mobility among the intellectual disciplines is a very important factor in producing innovation in the application of science.

Closely related to this and perhaps a sub-aspect is that very often the application of science occurs through the movement of people from basic science into technology. A look at the industries which have been most successful in producing technological innovation in an organized way—and I think this is particularly true of the communications industry, the information processing industry, and the chemical industry—reveals a very significant pattern. People came into the industry as basic scientists, trained in the latest techniques of basic research at the frontiers of scientific knowledge, but gradually they began moving up through the business hierarchy itself, which thus became infiltrated with the points of view, the techniques, and the knowledge of the latest basic science.

I think that if there is one factor that is more important than any other in the translation of science into technology, it is the gradual translation and diffusion of *people* from science into technology. Perhaps a corollary of that is my feeling that one of the essential conditions of continuing technological innovation in this society is the need for a certain degree of overproduction of scientists. Dr. Rabi made this point very nicely in his talk when he said that the thing that really revolutionized the chemical industry in Germany was that the Germans trained too many chemists. There being a lot of chemists without jobs, they went out and started a new industry.

A number of people have pointed to the fact that in recent years, outside of the military field, there has been a lag in technological innovation in this country. Part of the reason for this has been an underproduction of scientists. As a result, people have not been forced to jump fields as much as they might have. I think this is a situation that is probably going to right itself in the next ten years. During the 1950s we had a very peculiar situation. In the educational system we were dealing with the low birth rate of the 1930s at the very time that the demands for scientifically trained people were at their highest. Thus, in the

1950s there was a severe dip in the number of people graduating with bachelors' degrees in science.

In the 1960s this situation is going to be reversed. In fact, there is going to be a sharp increase in the number of people in the college age groups, so that I think there will tend to be, in some sense, an overproduction of technically trained people. This means that there will be more and more people with a technical background, a technical interest, and a technical point of view diffusing out into the whole society and in turn influencing the demand of society for technology.

This question of the movement of people from basic science into technology has several aspects. Of the Ph.D.'s in basic science, only about 20 percent end up doing research in what would be called basic science. The other 80 percent go on into various applied areas or into other areas altogether, although the number that go on into other areas at the present time is quite small.

Then, of course, in the industrial organizations there is the continuing process which I described, the movement of people out of the research laboratories into the operating or development divisions of the company.

Another factor which is very important and which, I think, has been badly neglected in recent years is the question of upgrading the teaching process at all levels in the light of modern scholarship. Dr. Rabi pointed to the fact that by and large modern science has remained the property of an elite and that we have been rather unsuccessful in diffusing this into the society as a whole.

I think one of the reasons for this is we have not gone far enough back into the educational process. We have regarded advanced training in science as something on which we could build a more or less static and fixed body of knowledge—science and mathematics taught the way it was fifty years ago. Today there is a growing realization that every new discovery at the frontier of knowledge has profound implications for the way children are taught, even at the kindergarten level. In other words, the educational process is an integrated process, and every time scholarship advances in any one field—not only in science—it has an effect on all other fields. Therefore, scholars have to worry a great deal more about the implications of their scholarship for the whole educational process.

The relation to technology and to technological innovation is simply this: if we are going to realize the potential of science, we are going to have to have a much more scientifically oriented education, not only for people who are going to become scientists, but also for people who are going to become something else. To some extent this has already occurred under our noses, perhaps without our realizing it. If you talk to someone who graduated from college in 1905 or 1910, you will find that he then took math courses which are now taught in the sixth or seventh grade. In the summer of 1963, a group in Cambridge, Massachusetts, attempted to look ahead twenty years, setting goals for the teaching of mathematics from kindergarten to the twelfth grade. The group included not only professional mathematicians, but also people in the education system at all levels, from elementary to high school. This group concluded, not with certainty but at least with some degree of conviction, that within the next twenty years the average high-school student would have a mathematical education equivalent to that of the average science major in college today by the end of his senior year. Perhaps this is an exaggerated point of view; nonetheless, it represents a goal which I think must eventually be attained. I don't think this will come about by increasing the amount of time that people spend in these fields, because we simply can't afford more time. It will come about by being more selective, more organized, and more careful about the context of what is taught and about the attitudes that one wants to inculcate.

What I am suggesting, then, is that this movement in education, which I hope we've just seen the beginning of, is really one of the most important aspects of the translation in the future of basic science into technology.

There is one other topic that I would like to say something about. It has to do with the dilemma of the government in trying to assess and control scientific priority. The problem which I'm trying to pose is this: To what degree, and why, is it a government problem?

As you know, two-thirds of our research-and-development funds come from the federal government. It is clear, then, that the way in which the federal government chooses to allocate its resources in the technological area has a profound influence. Of course, I think it is important to point out that of this very large sum of $15 billion that is sometimes conveniently labeled "for science," only about 7 or 8 percent is for what a scientist would call science. The other 92 percent is for technology. Still, we're talking about science and technology, and certainly federal funds even for the latter are very important.

Now, the first thing that must be kept in mind is that roughly 91 or 92 percent of this allocation is for space, defense, atomic energy, and other fields related to national security. Perhaps another 5 percent is for health. This leaves only about 3 or 4 percent for all other applications of science.

Secondly, basic science is, in a sense, the infrastructure or the social overhead which supports all this expenditure for the application of science, including education. I don't think one can really make much of a separation between education and research, because, at least at the most advanced levels, spending money in certain fields of research means that you are educating Ph.D.'s in those fields. The way you spend money in the universities for research is also the way you educate people at the most advanced levels.

Now the question is: to what extent can or should the government attempt to control this resource allocation? Basically, the allocation is politically controlled, and that, in my opinion, is as it should be. A democratic society has to determine its own goals, but the problem, of course, unlike many areas in which the government spends money, is that the relationship of the activities to the goal is often unclear. The more basic the science, the more unclear the relationship. The question is: how do we arrive at a judgment in such matters?

One might say that in the allocation of resources for science and technology, one has to look at the problem as a two-dimensional grid. Along one dimension one should place priority of social goals, and along the other dimension one should place priority of intellectual opportunity. In my opinion, if you look at only one dimension you are in trouble. If you allocate your resources only in accordance with the practical importance of a problem you don't solve the problem. On the other hand, if you allocate your resources only in accordance with the opportunity of making progress, with the opportunity for advancing knowledge regardless of the importance of the problem to society, you also don't make the maximum progress. So there is some kind of an optimization involved in this two dimensional grid between opportunity and need. It is this, basically, which makes the problem of resource allocation in science so much more difficult than it is in other fields, because everyone is going to have a somewhat different perspective and a somewhat different view.

Let me give you some illustrations. I can think of two cases where resources have been allocated on a fairly massive scale, largely on the basis of need rather than on the basis of opportunity.

The first example is cancer chemotherapy. This is an area where clearly the stakes were tremendously high. The discovery of a chemical that would produce a cure for cancer or for certain varieties of cancer would be a most welcome thing. The National Institute of Health, under pressure from Congress, in turn under pressure from certain elements of the medical community, launched a massive program of research in cancer chemotherapy. Basically, it was an empirical program for trying to find a drug that would have some influence on cancer. It consisted of trying as many different possible chemical compounds with as much guidance from science as was available. I think one would have to acknowledge that this program was unsuccessful. This, I would say, is an example of the direct attack on a recognized social problem.

Another example of the direct attack is the initial attempt to get control of thermonuclear power. There was a period in the middle 1950s when a group of scientists and politicians felt that the prestige value of accomplishing such a feat would be tremendous, and so a large amount of money was devoted to trying to build devices which would control thermonuclear reactions. The trouble was that the program was unable to make much progress because there was not enough basic knowledge. There were too many different ways one could go about doing it, and there was no basis for deciding which way was best.

But this is a matter of judgment, because to take an illustration on the other side, it is clear that the decision to go all-out in the military field for an intercontinental ballistic missile was probably delayed unnecessarily. In other words, the state of technology was such that this could have been started somewhat earlier. The same could be said about satellite technology.

Those who argue that if you want to do something you should just go ahead and do it can find plenty of examples to support their point of view. And those who say that if you want to do something you should first develop the basic knowledge you need to go ahead can also find plenty of examples to support their point of view. Therefore, in the last analysis, we come back to the subjective judgment—perhaps to the instincts, the intuition, of wise people.

But this is, at any rate, an indication of the kind of dilemma you face in trying to allocate scientific resources. Basically it's the dilemma that nobody really knows which science is going to be relevant to the particular goals you are seeking. The natural scientist doesn't know, and the political scientist doesn't know. The only mechanism I've been able to discover for arriving at these judgments is the good old-fashioned method of arguing. As many different kinds of people have to be brought into the argument as possible. This is the way it is really being done in practice, now, in Washington.

The President's Science Advisory Committee has strenuously resisted being put in a position of making the decision as to scientific priorities. Basically, these priorities are still determined by the Bureau of the Budget, *de facto*, with as much advice from the scientific community as they can get, again depending on the particular area. The Defense Department, for example, has made extraordinary strides in recent years in developing methods for evaluating the cost effectiveness at least of development; however, this kind of cost effectiveness evaluation[1] has not been developed for basic research, nor has it been developed for most areas of research outside the defense field.

DISCUSSION

TECHNOLOGICAL CHANGE, THE MARKET PLACE, AND THE SUPPLY OF SCIENTISTS

Comment: I would like to interject another point of view. Quite apart from science, technology operates in the market place, and this is the thing that makes it go or not go.

Irving Langmuir received a Nobel Prize essentially for developing the electric light bulb and, incidentally, making a great deal of money for General Electric, but the man who probed the market was Thomas Edison, and that was his function. Whatever you do, technology has no viability unless it survives in the market place.

[1] The use of statistical techniques to compare the imputed benefits of a particular expenditure with the imputed costs.

When you talk about technology in the abstract and relate it to science, without taking into consideration the market place, you're not talking very seriously. As far as transistors are concerned, they were made possible only because the government was prepared to underwrite the probing of the market place.

Mr. Brooks: I certainly don't disagree with that. The only sort of footnote I would add is that the market place may not be the whole thing.

Comment: Let me give you an illustration of why I think you can't rely wholly on the private market place. We economists in government worked over a period of time developing the use of input-output tables.[2] Along came a change of administration in 1953, and this important development was cut off. Fortunately the project was reinstituted in 1961, so we are in business again. In the interim some of my men took a leave of absence to see if in some way or other this work could be continued. A prominent financier was even interested in trying to develop this concept purely on a private initiative basis. However, these efforts all collapsed. The point I'm making is that even though the private enterprise system has more imagination, more freedom of action than we in government do, it isn't always able to meet these kinds of social needs.

Mr. Brooks: This was really the point I was trying to make before, that is, it isn't clear to me that one can apply the test of the market place to these kinds of products. I think one has a clear sense that just because the thing doesn't stand up in the market place in the conventional sense, it doesn't mean that the special benefits don't exceed the cost.

Comment: As one of the sociologists present, I want to question this emphasis that has been placed on the market place by some of tonight's participants. Now, the market place is not simply private individuals willing to buy or not buy products, which seems to be the old-fashioned notion. It's a very much

---

[2] Tables used by economists to ascertain the actual physical allocation of resources required by alternate economic policies.

more complicated thing. The market place is a combination of private individuals, government acting as a single individual, and various other collective groups.

The really important question is: how does one organize a market place? Quite clearly one can build a monorail, let's say from Boston to Washington, which would speed people along in just about the same time it takes now to go by plane, but one would have to buck a tremendous lobby consisting of the gasoline people, the automobile people, and others who are able to get appropriations from Congress in order to subsidize roads.

Comment: I agree completely. What Edison actually did was to organize—or as I would say—to probe the market place by stringing wires across rooftops and violating laws.

Yet this probing of a market can be a complicated business. Let me give you some examples of why this is so. Robert Mac-Namara, when he first came to the Ford Company, made one of the greatest mistakes in automobile history in probing the market with the Edsel. American Telephone and Telegraph made almost as bad a mistake, in reverse, in probing the market place with its coaxial cables across the Atlantic. Because they probed it in a narrow, unsophisticated way, they completely underestimated the potential use.

Comment: I was interested by your remark that technology is spread by the overproduction of scientists. Is it really overproduction or is it that scientists become "burned out," because the field grows faster than they do, and they therefore move into other areas?

Mr. Brooks: I really don't know. We all have opinions about how science is translated into technology. Personally, I think that it occurs for a variety of reasons, and I don't think we can pick any one reason. Certainly I think there is some validity to what Dr. Rabi says, that if there are fewer opportunities in basic science than there are people to do basic science, this will cause some of them to try to develop and exploit the potential technology inherent in present-day scientific knowledge. I think that what we really want to do is to overproduce people who have more skill than is necessary for the economy, rather than people with less skill than is necessary for the economy.

Comment: I don't quite see this point. I believe that in Europe there has been at various times an overproduction of trained personnel at all levels of knowledge with the result that many individuals have been compelled to work at jobs that were below their scientific competence. Still, there hasn't been any significant increase in the number of inventions or technological innovations. The real reason scientists go into industry is that they can earn a much better salary if they do.

Mr. Brooks: I think by and large, in terms of the civilian economy, Europe has been more inventive in the last ten years than the United States.

Comment: Just having a surplus of scientists, it seems to me, is not enough. It depends very much on whether there is a market for the services of those surplus scientists. The surplus of scientists in Germany in the nineteenth century was absorbed by the German chemical industry. I don't think they would have been absorbed by the American chemical industry at that time.

Fifty percent of all research funds spent in this country is devoted to improving that 10 percent of our national product which is expended on military purposes. Another 40 percent is devoted to that 20 percent of our national product which is closely related to the military effort. What bothers me is that only 10 percent of all research funds is devoted to the 70 percent of our national product which is absorbed by the civilian economy. Even if we were to create a surplus of people proficient in ultrasonics, they wouldn't improve the productivity of my corner dry cleaner.

Mr. Brooks: If anybody interpreted what I said to mean that all we needed to do to make the economy grow is to create a surplus of scientists I want to correct that impression.

Comment: What happened in Germany, as far as I know, is that the Germans wished to increase their agricultural production. A political decision was then made to devote considerable resources to that goal, including the provision of scientifically trained personnel. So it is not really true that there were numerous chemists in Germany without anything to do who suddenly created a chemical industry. It was the by-product of a very conscious political decision.

INTELLECTUAL OPPORTUNITY AND SOCIAL GOALS

Question: With reference to your two-dimensional grid of intellectual opportunity and social goals, I wonder if you would spell out what you meant by intellectual opportunity in cases where there is no apparent social goal.

Mr. Brooks: The government has been persuaded to make rather large resource allocations to high-energy physics in which the hope of practical application, at least in terms of immediate relevance, is pretty remote or nonexistent. Now, of course, this is a complicated thing because while high-energy physics has no immediate relevance as an intellectual discipline to anything practical, nevertheless, the by-products of working in high-energy physics have had tremendous technological implications, though I don't think this was realized when the decision was made to support this area of research. I think one can make a very strong case, although there are many people who would disagree, that the investment in high-energy physics has been an extremely profitable one, not so much in terms of the high-energy physics results, but more from the fact that we have had to develop Klystrons and half a dozen other things which became technologically important in other fields.

Comment: There is a school of opinion which holds that it always pays to make the investment whenever the intellectual opportunity is present.

Mr. Brooks: I think it depends on the magnitude of the intellectual opportunity, and also on the cost. Again, I think it's very difficult to put this in quantitative terms. One of the disadvantages of cost effectiveness analysis is, I think, that it always tends to penalize new technology.

Comment: I would like to pick up your statements about cost effectiveness and how it might penalize new technology. Cost effectiveness analysis is a form of utilitarianism, and if you once abandon utilitarianism you're in trouble. Of course, utilitarianism can encompass many different viewpoints. But the more explicit one is about the utilitarian calculations, the more explicit one is about underlying values.

Cost effectiveness analysis in itself is a technical innovation which should, I feel, be exploited more fully. The government, for example, by employing such a technique, could make clear what values it was basing its decisions on, thus affording people a more rational choice in the political realm. Federal appropriations might then no longer have the aspect of being a mere "pork barrel" operation. As long as the values and premises underlying the analysis are clear, there is no reason to fear cost effectiveness analysis *per se*.

Mr. Brooks: Perhaps I should have qualified my statement that cost effectiveness analysis can penalize new technology. What I meant is that cost effectiveness—as currently applied— can penalize new technology.

All I'm saying is that this particular method is overquantified. In using it, we tend to focus attention on those things we can measure and sweep aside those things that we can't, and therefore probably do not assign the proper weights to the various factors in our analysis.

Comment: I would like to say that as a matter of practice, cost effectiveness analysis might work out somewhat better than you have suggested. It occurs to me that the Atomic Energy Commission, in deciding to support high-energy physics, might say: "We don't expect any substantive payoff from this, and the various technological advances, their by-products, may not justify the expense either. But we do know, and this could be cranked into the cost effectiveness calculations, that we've got to gain the allegiance of the scientific community, and so the funds devoted to high-energy physics may be a small price to pay." In this way, we could allow for the intuition, enthusiasm, and intellectual acumen of individual scientists and other factors. I don't see any problem in incorporating these variables into your analysis.

Mr. Brooks: I think perhaps it could be done, but I think it's difficult. How do you quantify the elements you're talking about?

Comment: You figure out what price our society would have to pay for losing the loyalty of the scientific community.

Mr. Brooks: But who's to judge, in fact, that any particular decision would buy the loyalty of the scientific community?

Comment: I still insist that there is a rational basis for deciding whether to support your category of intellectual opportunity with additional funds. But there are limits, as you say, to how precisely this can be done.

Mr. Brooks: Intellectual opportunity does receive some financial support. I don't think I was arguing that it doesn't. Nevertheless it is much harder to evaluate the pay-off in basic research on a cost effectiveness basis than it is to decide, for example, whether to go ahead and develop a supersonic transport. I'm just saying that the criterion of cost effectiveness is much harder to apply in the case of pure science than it is in the case of development. Of course, that doesn't mean we don't apply it, only that we do it qualitatively, as best we can, rather than quantitatively.

Comment: Without experimental data, we are apt to make stupendous errors, and that is the point. Most of the time we don't have the data, which means we have to rely on intuition. In the case of the jet transport, every single person connected with the airline industry, including the manufacturers, underestimated the real return from it. If we try to rely on cost effectiveness analysis without someone being willing to gamble, nothing will ever be built.

Mr. Brooks: I can do a beautiful piece of operations research, and I can demonstrate conclusively that we should never have substituted steam for sail.

Comment: This proves that all of these decisions are essentially political. Mussolini decided to build roads when there were no automobiles in Italy; that was a pure political decision. Today we're engaged in making the same kinds of decisions. And the people who must make them have little more to go on than faith.

Comment: I'd like you to elaborate on some of your earlier remarks as to the political decisions that are made on the allocation of resources. I think basically you're in favor of going back to the old-fashioned method of arguing. But I'm wondering who should do the arguing. It seems that by default we are letting the Bureau of the Budget make these decisions, without the sort of debate you would recommend.

Question: Much of what you said in your opening remarks dealt by and large with democratic societies. How does your theory fit a totalitarian system? Do things happen in quite the same way?

Mr. Brooks: I think I would subscribe to Dr. Rabi's judgment that, on the record, totalitarian societies have been less successful overall in the application of science; that is to say, they've achieved fewer results despite greater effort.

Question: How do you judge that? Aren't you simply judging by your own standards?

Mr. Brooks: Of course, it's very hard to make an overall judgment, and it may be true that I'm setting my own standards.

Question: Let me help you by asking a question. Where do you expect the next most efficient automobile engine to come from, Russia or England, France or the United States? Where do you expect the next important synthetic fiber to come from?

Mr. Brooks: I think a totalitarian society can misallocate resources more efficiently than a democratic society can.

THE DIFFUSION OF SCIENTISTS AND SCIENTIFIC JUDGMENT

Question: You talked about the diffusion of 80 percent of the people starting out in science into fields other than pure science. I wonder if you have a judgment about the competence, from the point of view of scientific judgment, of the people who land in the higher reaches of American business where there is a connection between science and technology. I mean, in particular, the chemical, communications, and similar industries.

Mr. Brooks: I think this varies a great deal, but I think in the more technological industries, the level of scientific competence is very high.

Question: You haven't been bothered by the fact that it usually takes thirty years for a man to work himself up into a vice-presidency and by that time quite a few things might have happened to the science he was once taught?

Mr. Brooks: No. By and large, the man who has to make decisions doesn't have to understand all the details.

Comment: In judging competence in the university there is at least an outside evaluation system. American Telephone and Telegraph or Du Pont, on the other hand, are heavily dependent on inside information flows. It is very hard to be certain that these flows are related to reality.

Mr. Brooks: I wouldn't say that the universities did better or worse than industry in this respect. I don't have any general feeling about this at all. When you made your statement, I could think of good examples and bad examples, and I find myself unable to strike a balance.

Question: By and large, American corporations have built up their senior staffs completely from inside. On the other hand, the larger universities operate in almost a national and perhaps even an international market in securing personnel. Does this have some relevance?

Mr. Brooks: This is something that you know a great deal more about than I do. What are the actual facts about mobility in the top management level?

Comment: It depends a great deal on the industry. In advertising, for example, the rate of turnover is something like once every two or three years. In the railroads and the other basic industries the turnover is practically nil.

Comment: Aren't the industries using the same people that the universities are using? What about those strings of consultants?

SCIENCE, TECHNOLOGY, AND SOCIAL CHANGE

Question: I wonder if you care to direct some remarks to the social adjustments involved in the diffusion of science and the spread of technology.

Mr. Brooks: I think science and technology have been a very important indirect source of institutional innovation. For example, the new groups that have been set up to support research and development over the last fifteen or twenty years are among the most revolutionary institutional changes in our political structure. This has occurred more or less by accident.

My problem with your question is that I'm not quite clear in what context it's being presented. Are you asking to what degree can we use science and technology to instigate institutional change, are you asking to what degree have science and technology produced institutional change, or are you asking how can we control the institutional change that science and technology are producing?

Question: My question is quite simple. What would you say is the relation between science and technology on the one hand, and social change on the other?

Mr. Brooks: Nothing, except what I've just said. It's clear that they produce change and maybe we ought to worry more about controlling and directing this change—perhaps even anticipating it more than controlling it.

Comment: One can see that social adjustments have been greatest in the areas touched on by the weapons establishment, which has deeply involved itself in the process of diffusing science.

Mr. Brooks: It seems to me that many of the institutional devices which have been successful in the development of weapons systems have relevance, though they may not be immediately translatable, to other areas. The idea of organizing interdisciplinary groups to bring about innovation is a social invention that stems largely from weapons development. A very mundane example is the so-called summer study group. This same institutional mechanism, it seems to me, is being applied very effectively now in education, as well as to such problems as the decay of central cities. It might also have relevance to some of the other major sociological and economic problems facing the country.

Comment: But it might also be as profitable as medieval scholasticism.

Mr. Brooks: We've got to start somewhere, and, it seems to me, in some cases this might be a good place to start.

Comment: Contrasting what has happened in the defense program with what has happened in the effort to direct and control social innovation in the so-called underdeveloped nations is revealing. One can see the steel mills, the railroad systems, or the other physical signs of change, but not very much in the way of social change.

Mr. Brooks: I think that's because we really don't understand enough about social change.

Comment: Our discussion so far has been on how we can use science to improve technology. It seems to me that we ought to have a corollary: How can we use this improved technology to serve human beings? The object of improved technology, it seems to me, ought to be to improve human life, not to degrade it. I use these terms deliberately because when a change in technology destroys the skills of thousands of people, it degrades their humanity.

It seems to me we've got to give some thought to the question of how rapidly we should make equipment obsolete. I know of certain procedures in my field which would lead to great savings but which would involve a great social cost. How do we balance these two? It is particularly difficult since the antitrust laws prevent people from getting together to talk in a sensible way about these problems.

Comment: I have another question. Is more technology good for the mental health of the scientists, the technicians, and the society in general? Now, my personal life in the last twenty-five years, as a result of the improvement in science and technology, has I think, actually deteriorated slightly. At least in terms of the services I was once able to command and the less hurried pace at which I was able to enjoy them, I feel the quality of life has deteriorated.

Comment: The problem is that there are more people now claiming services. That is why you feel that you are being crowded out.

Comment: The heart of the question still seems to be, could part of the problem be an overly rapid rate of technological change?

Comment: I would like to say one word on this question of an overly rapid rate of technological change. The feeling is in direct proportion to one's age. I'm fifty-seven, and it seems *much* too fast. You're somewhere around forty, and it probably seems rather fast. But I can guarantee you that my own son and the boys I work with at Columbia don't think it is fast at all, so that in one word, what you are saying is, "I am getting old."

THE EDUCATION OF ENGINEERS

Question: You touched on the dilemma of engineering education and I would appreciate your expanding on that. The fact that Russia is producing so many more engineers than we are may be relevant here.

Mr. Brooks: The dilemma of engineering education can be best expressed by saying, "What is an engineer?" The dilemma has been produced by the increasing role of science in engineering. I think the comparison between Russia and the United States, to answer your specific question, is really almost meaningless, because what is meant by an engineer in Soviet society is very different from what is meant by an engineer in the American system. In other words, engineers in the Soviet Union fulfill many of the same functions that other professionals in this country, such as economists, lawyers, and business executives, perform.

I think it's more meaningful to look at the total percentage of the age group which is educated to a given level than it is to take a particular category such as engineering and ask how many such persons are being produced. However, it certainly is true that Soviet society has decided to allocate most of its educational resources to the production of technically trained people, whereas we have not.

One of the things that is surprising about this subject is the great stability in the percentage of technically trained people that we produce in America, despite all the social perturbations in the last fifty years. The division between those going into the social sciences and the humanities on the one hand, and those going into the sciences and engineering on the other, whether you talk about it at the B.A. or the Ph.D. level, has remained almost constant. The same thing seems to be true of the Soviet Union, although the division is different.

But to take up your question about the dilemma of engineering education, I think basically the dilemma is concerned with the following questions: What is an engineer? How much of his training should be science? Is there a discipline called engineering which is separate from science, or is it only a particular collection of sciences which have a different relevance? How much science and how much mathematics should there be in the education of the engineer as compared with other subjects? By and large, what is the function of the engineer in relation to the scientist? Is it better to simply train more scientists and convert them into engineers later on? What is the relationship between the practicing engineer and the engineer who does research and development?

Comment: I don't see any dilemma and, in fact, I've been wondering for a while now what we are talking about. This problem of allocation is a perpetual one. Every society has decided, on the basis of certain tenets, how to allocate its resources. If it is a religious society, there will be more priests than poets; if it is a military society, there will be more warriors than scientists. In my mind there is nothing new about the problem—it is a question of what the tenets of the society are, and the tenets of a society are determined by the pressures of various groups, whether they are commissars or the owners of private enterprise.

This idea of putting together people to discuss major problems is also as old as the hills. It was done in the Middle Ages when academies were founded all over Europe and social scientists met with mathematicians, with engineers, and with philosophers. There is no dilemma in engineering education. How much math? More!

Some engineers have become good scientists, but I don't know of any good scientists who have become good practicing engineers.

Comment: We seem to consider science and technology as unitary or monistic activities. Actually, both are whole congeries of activities. There are different kinds of science, not all of which lend themselves to technology. There are different kinds of technology, some of which are applied science, others of which are purely empirical. Because we haven't made any clear distinctions there is, I think, a great deal of confusion. Someone, perhaps, was impelled by the original scientific discovery, but others who came later were simply impelled by the internal dynamics of technology, making improvements in devices, gadgets, or procedures without any knowledge of the basic science involved.

My second point deals with the assumption of a linear relationship between science and technology. Others have already pointed out that there is also an economic factor which enters into technology, and I simply want to add that there is more than an economic factor. There are, in fact, many social and cultural factors.

Government, Education, and Civilian Technology

by ALVIN M. WEINBERG

Director, Oak Ridge National Laboratory

EVER SINCE John Kenneth Galbraith's *The Affluent Society* appeared, it has been popular to say that civilian technology in the United States is lagging. One of Galbraith's main contentions is that the fragmented industries, like textiles, coal, or housing, cannot support the research and development necessary to keep them competitive. For such industries the government ought to take responsibility for the relevant research and development and in this way help overcome the technological lag.

I shall first illustrate Galbraith's point by describing how atomic energy, a highly unified industry enjoying much support from the government, seems to have overtaken coal, a fragmented industry enjoying little government support. Second, I want to discuss three obstacles to the application of modern science to the civilian economy: the remoteness of the university, the trend toward purism in our technical education, and our preoccupation with military technology. Finally, I shall suggest a scheme by which the government can help our fragmented industries overcome their technological backwardness.

THE NEW SITUATION IN ATOMIC ENERGY AND IN COAL

Ever since atomic energy was discovered people have been saying that ten more years were needed for it to become economical. Actually in the past few months atomic energy has become economically competitive with energy from coal in half the United States. This is perhaps the most important happening in atomic energy since the invention of the H-bomb, yet it has largely escaped public attention.

This achievement has created a sensation in the atomic energy community, and has spread consternation among the coal people. Not only was this development overlooked by the public at large; it had not been anticipated even by the distinguished former Chairman of the Atomic Energy Commission, David Lilienthal, who recently implied that atomic energy would not, at least in our lifetime, become economically competitive.

The event which marks atomic energy's becoming competitive occurred in February, 1964, when the General Electric Company agreed to build for the Jersey Central Power and Light Company at Oyster Creek in New Jersey a boiling water reactor that is estimated to produce electricity at rather less than 4 mills per kilowatt-hour. For comparison the going rate for the generation of electricity in the New York area is between 6 and 7 mills. The newest conventional plants are coming in at around 4 mills, whereas this atomic plant is estimated by the Jersey Central Company, its buyer, to be able to produce electricity at about 3.8 mills per kilowatt-hour, for the first five years of operation, and at 3.5 mills per kilowatt-hour for the next ten years of operation. More impressive than the overall cost per kilowatt-hour (the overall cost depends sensitively on the fixed charges, which, in this case, are abnormally low) is the unit capital cost of the reactor. If the reactor achieves an output of 620 megawatts as General Electric believes it will (the nameplate rating on the plant is 550 megawatts), then the capital cost per kilowatt installed will be $113. This is cheaper than the installed cost for many of the most modern conventional plants.

This major advance cannot be attributed to any highly sophisticated application of new principles to the generation of atomic energy. It seems to have been the result of two factors. The first is the intense competition between the General Electric Company and Westinghouse, the two major reactor suppliers in this country. Westinghouse had been winning the majority of recent reactor contracts, and General Electric apparently lowered its prices to compete with Westinghouse. However, I have been told that even at this low price General Electric will make money if it can sell from three to six reactors a year.

The second factor leading to this advance is the development of large pressure vessels to the point where one reactor, housed in a single vessel, can produce 600 megawatts of electricity. Reactors as large as this are cheaper, per unit of output, than are smaller reactors.

The combined effect of these two factors, one of which is technical, the other not, is to make very large atomic energy plants competitive in half the United States. This means that atomic energy will probably be competitive in much of the rest of the world.

Shortly after the General Electric bombshell exploded, I received a letter from one of our country's largest coal operators. He brought up Galbraith's point—that the coal industry is too highly fragmented to do the kind of research it needs, and that atomic energy, with its marvelous research facilities paid for by the United States government, offers coal unfair competition. He asked whether it would be possible to direct at least some of the effort of our government-owned establishment to research on coal.

I mention these incidents to illustrate several points in this general discussion on civilian technology. The first is that we must be very humble in making predictions. Even the Atomic Energy Commission was very pessimistic in its timetable. It expected atomic energy would be generated at 3.5 mills per kilowatt-hour not earlier than 1975, some seven years later than it now appears we will be able to achieve this figure.

The second point is that the fragmented industries do lag behind the integrated industries, as well as behind the industries in which research is government-supported. Whether the primary trouble in coal is too little research is by no means clear. After all, competition between Westinghouse and General Electric, no less than research on pressure vessels, brought us economical atomic energy.

To those of us who have worked for twenty years to make atomic energy competitive with coal, the General Electric announcement has been sweet wine. On the other hand, now that the happy day is here, the wine tastes bitter. I live in a coal-mining area. In Anderson County, Tennessee, where the Oak Ridge National Laboratory is located, five thousand coal-mining people are now living on surplus food from the government. The economic triumph of civilian atomic energy means that many thousands of coal miners will lose their jobs.

OBSTACLES TO ADVANCED TECHNOLOGY

What are the obstacles, as I see them, to the application of science to civilian technology? Is deficiency in technological research the main trouble with the fragmented industries? Can research really save textiles from Japanese competition, or coal from displacement by atomic energy? The usual view is that research, intelligently performed, can work miracles in the fragmented industries just as it has in the integrated industries. The obvious bars to effective research in the fragmented industries are insufficient money, lack of tradition in research, and incompetent management. I would like to discuss three other possible obstacles to effective research in civilian technology.

THE UNIVERSITIES

First, I ask, does research in civilian technology lag because of snobbish attitudes toward applied sciences in our universities?

This possibility has been raised recently by Edward Teller and by Hans Bethe. Edward Teller has said that in applied research, the United States suffers because our universities inoculate our best people with the idea that pure research is the one thing worthy of their attention.[1] Bethe strikes a similar note:

There is another kind of social responsibility which concerns the universities. I think the university ought to train students for all the many jobs in industry, in government, in laboratories, in the universities themselves. Now especially in the last ten years, the universities have emphasized the most advance training in science and engineering. This undoubtedly is good. However, for many of the socially important functions it is more important to have good training in some of the *less* advanced fields. A good design engineer is still needed as much as ever, and in aeronautics we still need engineers to build and improve planes rather than only engineers interested in magneto-hydrodynamics and space flight. In physics we have an obligation to give training in such subjects as atomic physics, which is by now an almost forgotten field. Research in physics has moved to high-energy problems and field theory, but for practical applications atomic physics is still the most important field, and the larger fraction of our students will be concerned with practical application. It is a difficult task for the university professor to dissociate his direct research interest from the subjects which are socially useful.

There is especially the question of establishing a sense of values in the student. It is very likely that the student will value most the things which their professors do. As a consequence, there are many frustrated scientists working in industry who think that they are not doing first-class work because they are not doing the same things that their professors did in the university. At the same time they are doing extremely useful work and they would do it better if they felt in their bones that this was important. So I think it is one of our social responsibilities to instill the right sense of values in our students.[2]

[1] "Hearings before the Select Committee on Government Research of the House of Representatives" (Washington, D.C., 1964), Part 2, p. 941.
[2] *SSRS Newsletter*, published by Society for Social Responsibility in Science (Gambier, Ohio), No. 128 (February, 1964), 4.

I generally agree with Bethe and Teller's view: a certain intellectual snobbishness has grown up within the universities, which creates a distaste in graduate students for practical work. From my work on committees in government I will go further and say that the attitude of the university, particularly the deification of research at the expense of engineering, colors much of the advice which the government gets for dealing with technological problems.[3]

Harvey Brooks has suggested one remedy: to overproduce basic scientists so that many of them will have no choice but to go into applied work. This may be practical advice, although I see difficulties. First, only the poorer scientists will go into practical work; and second, there will never be an overproduction of basic scientists as long as the money earmarked for basic science exceeds what the available scientists can spend. This is very much the case in many parts of basic science, perhaps most notably in high-energy physics.

Another, and perhaps more attractive possibility, is the establishment of schools of applied science within the universities. For these schools to succeed, the mother university must take them seriously. I mention this because so often one detects a reluctance on the part of the university president to accept applied science as a proper function of the university. This attitude may have merit, but in that case special universities in which applied science finds a congenial home will be needed.

THE NEW CURRICULA

The trend toward purism in the universities is spreading. It is invading our high-school curricula. I refer, of course, to the very extensive and in many ways admirable curriculum reforms such as the PSSC[4] in physics, the School Mathematics Study

---

[3] See my article, "The New Estate," in the *Bulletin of the Atomic Scientists,* XX (February, 1964), 16–19.
[4] Physical Sciences Study Committee organized at MIT in 1956 and supported

Group,[5] and similar enterprises in biology and chemistry. These reforms achieve two different purposes. First, they demand very much more of the student than do the previous curricula. With this I think all of us are in strongest sympathy. Second, they seem to inculcate in students a feeling for the modern state of the science that they are studying. With this aim I am only partially in sympathy.

The point of view of the modernist is generally narrowly specialized. The things that interest modern mathematicians are not usually the things that the mathematically literate scientist must know, not to speak of the mathematically literate engineer. In so far as the curricular reforms have been captured by the professionals of narrow outlook, by the same breed who in the universities are responsible for the deification and fragmentation of pure science and the denial of applied science, I believe the curricular reforms are dangerous. We shall get young people who have little sympathy for science outside their own speciality, and we shall get people who are less able to carry out applied research, since applied research so often is eclectic and requires a broader if less intensive knowledge than does basic research.

In the long run these developments will affect our capacity to perform applied research, and the matter is therefore germane to this issue of technological change. A recent review by the Britisher, L. S. Goddard, of *The USSR Olympiad Problem Book* [6] points up the danger. Dr. Goddard says:

During the last decade or so there has been a strong trend, particularly in America, to push back into the schools some of the abstract

by NSF, Ford Foundation, Alfred P. Sloan Foundation, Fund for Advancement of Education, and others.
[5] Headed by Professor Edward G. Begle of Stanford. For statement of objectives, see School Mathematics Study Group, *Newsletter No. 6* (New Haven, Yale University Press, March, 1961).
[6] D. O. Shklarsky, N. N. Chentzov, and I. M. Yaglom; rev. and ed. by I. Sussman; tr. by J. Makovich (San Francisco and London, W. H. Freeman, 1962).

notions of mathematics, together with the appropriate terminology. School children will have heard on many occasions of sets, operations, mappings, rings, fields, vector spaces, Boolean algebra, and so on. The fashion spreads and carries with it a social obligation to fall into line and follow the crowd. It is a dangerous experiment, and may lead to a generation of mathematically illiterate. The *Olympiad Problem Book* would suggest that the Russians are having none of this. The problems cover the most classical parts of mathematics, and seem to indicate that the best training for the young mind is still to come to grips with problems—some of them rather slippery—whose solution depends on semi-original applications of standard results in elementary mathematics. All teachers concerned with the 'new' thinking in mathematics should study the book carefully. [7]

The review gains much relevance from its juxtaposition to another review [8] in the same issue of *Endeavour* of a book dealing with the remarkable advances in nonlinear differential equations that have been made by the Russian mathematicians. Nonlinear differential equations are central to the application of mathematics to control systems.

Could it be that our professional mathematicians, in imposing on the coming generation their own narrow prejudices and fashions, are reducing our ability to cope with future technological problems, and that the Russians are not making this particular mistake? I think the evidence is pretty good that we are being caught up in just such shortsighted pedagogical strategy.

MILITARY RESEARCH

Additional obstacles to effective research in civilian technology are the government policies which, as stressed by Assistant Secretary of Commerce, J. H. Hollomon, funnel so much

[7] "Teaching of Mathematics," a review by L. S. Goddard, *Endeavour*, XXIII, 47 (January, 1964).
[8] "Process Control," a review by G. Temple, of "The Mathematical Theory of Optimal Processes," by L. S. Pontryagin, V. G. Boltyanskii, R. V. Gamkrelidze, and E. F. Mishchenko.

effort into military research and development and so little into civilian technological development. Is Japanese textile machinery better than American textile machinery simply because the Japanese do not spend much on military research and development, whereas we do?

Secretary Hollomon has urged the creation of a civilian technology extension service, similar to the Agricultural Extension Service, which would mobilize the universities for the benefit of small industries in much the same way that Land Grant Colleges have helped farmers. So far Hollomon's scheme has had tough sledding in Congress, partly because some industries, such as building, do not seem to want very much to be modernized.

Two possible ways out of the dilemma present themselves. The first is to try to improve the diffusion of technology, to make more of the by-products of our military research and development available for civilian use. My own view is that these by-products of military research and development tend to be exaggerated. Nevertheless there are important cases of such diffusion and it is worth noting some of these cases.

A beautiful example of a by-product of military technology is "Sketch Pad," which I saw recently at the Lincoln Laboratory in Cambridge, Massachusetts. "Sketch Pad" is a computer system which enables a draftsman to draft perfect engineering drawings freehand. It works as follows: suppose you are a draftsman and you want to draw a square. As you draw the sides of the square on a cathode ray screen with a "light" pen, the computer recognizes what you are trying to draw and trues up the square. Suppose now you want to draw the arc of a circle. You draw the circle freehand as best you can, and the computer converts it into a perfect circle, the *best* perfect circle that can go through the points that you are drawing. "Sketch Pad" will do much more. Suppose you wish to design an electronic circuit diagram containing capacitors, resistors, and inductors. As you sketch a rough symbol for a capacitor, out comes a perfect symbol for a capacitor, and so on for the other circuit elements. Moreover, when the diagram is finished, the computer will tell you the voltages and phases at every point in the system. I found "Sketch Pad" to be an almost incredibly ingenious bit of fallout from military research and development.

Another worthwhile example of such fallout is the application of the zonal centrifuge, developed at Oak Ridge for the separation of uranium isotopes, to the separation of the moieties within cells in biological systems. Biologists often must separate homogeneous particles from masses of cells. For example, one of the polio vaccines, which has been distributed to many hundreds of thousands of children, was found to contain the SV 40 virus, which induces cancer when injected into young hamsters. With the zonal centrifuge one can in principle clean the vaccine of any foreign viruses.

The transfer of the zonal centrifuge technology to civilian application was remarkably smooth. Nevertheless, at Oak Ridge we have been devising ways to encourage a broader transfer of government-sponsored development to civilian technology. We have established an Office of Industrial Cooperation, which actively seeks developments at our laboratory that may have commercial value, and then calls these developments to the attention of interested entrepreneurs. The alternative to such formalized transfer of information is the informal, and to my mind, unethical, if not illegal, transfer by individuals who work at a government-supported laboratory or a university during the weekdays and at their business on weekends. It is our hope that the Office of Industrial Cooperation will discourage such practices and yet transfer information effectively.

The Office has been in operation for less than a year; during this time about half a dozen specific pieces of technological fallout have been identified—solid-state counter, high-intensity ion sources, the zonal centrifuge, to mention a few. We have held several industrial participation conferences at which interested companies have learned about these developments, and I believe that in at least a few cases some of the techniques developed at our laboratory will become articles of commerce.

THE GOVERNMENT'S LEVERAGE AS A LARGE BUYER

However, in spite of our successes, I am still rather skeptical about our civilian industry relying in any major way on haphazard technological fallout. The technological distance between a rocket and a man's shirt is just too great to expect developments in the former to have very much bearing on the latter. I believe therefore that a scheme suggested by Joshua Lederberg of Stanford University and then elaborated by Robert A. Charpie of the Union Carbide Corporation,[9] may be much more effective in bringing the immense power of the government to bear on our civilian technology. Lederberg and Charpie point out that the government agencies in pursuing their regular missions exert enormous leverage on the civilian market place. For example, the Department of Defense undoubtedly is one of the largest single buyers of white shirts or raincoats, the General Service Administration, of typewriter ribbons, and the Federal Housing Authority is the largest financer of small homes. Why not use this purchasing power of the government as a lever to encourage technological innovation in the shirt industry, in the raincoat industry, in the typewriter ribbon industry, and in the housing industry? The Quartermaster Corps could support research on how to make a cheaper and better raincoat; I understand that it has recently come up with a remarkable waterproofing chemical for raincoats. Perhaps other rather unlikely areas of civilian technology especially in fragmented industries could be helped greatly by judicious and imaginative exercise of the government's power to buy common articles of commerce.

I have touched on several topics that relate to the question

[9] Private communication.

of how to accelerate technological innovation. In a more thorough discussion one must consider the connection between technological innovation and economic growth, between expenditures for research and technological innovation, and perhaps, most important, the sociological consequences of technological innovation. What should we really do about the coal miners in East Tennessee who lose their jobs when utility companies realize what a miracle the General Electric Company has wrought, or about the plumbers who will work less if FHA actually does rationalize the construction of small homes? Let us hope that forums like this by helping our technologists gain more insight than they now have into the social problems raised by their activities, will lead to effective solutions of these problems.

DISCUSSION

PURE AND APPLIED SCIENCE IN THE UNIVERSITY

Comment: In my mind it is dangerous to try to push the universities into applied science because you never know what developments in pure science will have valuable applications. For example, until about five years ago topology was an unknown word in engineering and only an extremely abstract mathematician knew about topology. Within the last five years topology has become absolutely essential in designing the frame of airplanes or of skyscrapers. Those of us who are interested in the latest studies of complex structures necessarily use topology. I do not know whether one can accuse the universities of overemphasizing purity. Indeed I believe that the universities ought to be as pure as possible.

Mr. Weinberg: I don't think that my complaint about the universities could be described accurately as a complaint against their emphasizing pure science. The function of the university is both the creation of new knowledge and the preservation and teaching of knowledge. It is not always clear to people that the creation of knowledge is consistent with the preservation of knowledge and that many of the areas of science which lose fashion in the university community are still very important. It is the neglect of these areas by the university that I am really against.

Comment: The Germans made a distinction between universities and *hochschule,* so you could have a high development of theoretical science in the university and still have very good applied science in the *hochschule.* We have this distinction to a slight degree here. MIT historically represented an institution that was useful to the country but different from a university. The British, who have very great scientists, have apparently lagged on the technological front. But this may not be due to the universities at all.

In the United States for a long time we had trouble getting enough highly trained engineers because our engineering group didn't have enough influence. Now we begin to hear the opposite side of the story. The big new flows of governmental funds permit all kinds of new fashions, new prestige, new influence centers to be developed.

On the other hand in the United States it is very easy to keep on getting a lot of prestige, and not a little money, for being "concerned with nothing but ideas." Incidentally, the same complaint is heard in Russia. Everybody is in theoretical physics and you can't get anybody to do any experimental work because the Russian government is using bad judgment in giving all the prizes to people who do theoretical work. Moreover it is inevitable that in a university environment people will always give prestige to ideas. The advances in theory are always going to be the center of the university.

Comment: University communities are extremely conservative. Yet the fact that many universities are changing their emphasis in science and engineering and have done it fairly rapidly would indicate that there is a great deal of flexibility in our educational institutions. Part of the reason engineering schools have changed is the evolving nature of science and technology. Science and technology have changed the nature of jobs, and we have responded by changing the nature of our education. It seems to me that the universities do have contact with reality, with the market place.

THE BAROQUE IN MODERN PHYSICS AND MATHEMATICS

Mr. Weinberg: The trend toward extreme abstraction in mathematics is an example of a "baroqueness" in science about which the great mathematician von Neumann complained before he died. This trend is most apparent in mathematics, but it exists in physics—that is, in those specialties which have lost contact with the rest of science and mainly are elaboration of details which are of little interest except to the most highly specialized experts in the field.

Questions: Isn't it just the rare few who pursue this baroque mathematics and physics?

Mr. Weinberg: Yes, and I don't object to any kind of science, no matter how baroque, as long as it doesn't demand a great deal of support from society and, more important, doesn't impose its extreme viewpoint on the whole field of scientific and technical thinking. Many of the mathematicians who have worked in what is called the unified tradition, associated with names like Poincare and von Neumann, objected so strongly about this tendency toward the baroque that they drew up a Mathematicians' Manifesto in which they urged that any new curriculum reforms preserve the unified tradition.[10] They felt that mathematics is too important for the curriculum to be entirely in the hands of the professional mathematicians.

Comment: I don't think that it is a question of whether applied or pure science ought to be taught. I think it is a ques-

[10] *American Mathematical Monthly*, LXIX (March, 1962), 189–93.

tion of how much contact with reality remains in pure science. We no longer feel any need to relate pure science to what is going on in the physical world. I suggest that we go back to the historical method of presenting pure science and pure mathematics, a method that began with reality.

Comment: The great power of science comes from abstraction. If you keep your feet on the ground you will never discover anything. When Fermi became a full professor of physics at the University of Rome, some physicists who attended his lectures said that Fermi's ideas were completely divorced from reality, and that they were sheer nonsense, both pedagogically and scientifically. They were of course completely wrong. I cannot overemphasize the importance of abstraction from reality in the development of any real scientific new thought.

Mr. Weinberg: The problem goes deeper than this. You have pointed out that the physics that Fermi was expounding in 1925 appeared to some physicists to be a baroque manifestation of physical thinking that had no connection with reality. But other people realized at the time that it was enormously important because it really dealt with concrete objects. For example, I have argued that high-energy physics has some baroque elements; what I mean by this is that it deals with phenomena that are very far from our intimate and daily involvement with ordinary matter under ordinary conditions, and therefore bears little upon the rest of science. It is impossible for us to know today who is right. It may turn out that phenomena that take place in 10^{-22} seconds will be enormously important in the next twenty years. I merely think it to be much more likely that phenomena occurring in times of an atomic transition (10^{-12} seconds) will be more important as far as we human beings are concerned. I approve of people studying phenomena that take place in 10^{-22} seconds, if society can support such studies without neglecting other possibly more relevant matters.

Comment: It seems there should be a selectivity process

somewhere along the line where people who can cope with what you call the baroque can get into it, but where others who can't or don't want to can be offered some other type of approach to problems. As it is now we are feeding one type of thing to a mixed group, some of whom are capable of making the kind of discoveries you are talking about and others of whom just get lost.

SYSTEMS TECHNOLOGY, MOTIVATIONS, AND SOCIAL COST

Mr. Weinberg: Components technology has been more successful than has the much more complicated systems technology. For example, the components of a house are highly developed, especially since companies like Johns-Mansville have their own very effective research organizations developing new gadgets and materials for houses. On the other hand, we are much less efficient in putting the components together to make a house. Robert Charpie's suggestion that the government use its leverage in controlling loans for housing to encourage experimentation, say in organizing the building of a house, might be used to advantage here.

Comment: You must synthesize different elements when you attack problems such as these. In this instance you have scientists making recommendations which go right to the heart of our economic and social system and our political structure without reference to the existing society and its traditions. The way our society makes some of its decisions has changed very radically in the last twenty years. Perhaps this process of change is too important to leave as much in the hands of scientists as we have permitted. The intellectual energy inducing change in our society today rests in a fairly small community of scientists and related people which is not connected with our traditional means of introducing social change.

This concentration of intellectual energy has run parallel with the size of the controllable phenomena. In other words we deal in superb fashion with atomic energy, with computers and electronics, communication, space, and so forth. And this whole set of fields has the property that the problems you attack are capable of being confined to a small scientific area where you can concentrate intelligence upon it. If you look at the kinds of problems that were raised in Secretary Hollomon's paper—better housing, better road systems—technology is not going forward but is retrogressing. I submit also that there has been an implicit intellectual choice of the scientific and university community to abrogate their responsibility for the changes which are taking place in the social institutions. I think these are the kinds of problems which the seminar must deal with. This is where you really need the philosopher, because you are dealing with choice, you need the economist because you are dealing with the flow of human activity, you need the sociologist because you are dealing with the effect on the ways people live.

Comment: Society can be mobilized for particular ends very quickly. When you deal with systems you deal with interrelations, a very much more complicated area, and one you cannot manipulate as quickly.

The political process is the most elaborate system possible. It is quite clear that you can have now a pure abstract systems approach to the buildings industry, but you also have a hundred thousand builders, each of whom builds some ten houses a year. It is very hard to displace a hundred thousand builders on the basis of an abstract system, even though there might be a great abstract social gain.

We have had a very ruthless kind of economic democracy which is called the market. Sometimes it has been so ruthless that everybody tries to protect himself against its ravages. And the problem is to find some mechanism by which the political process can mesh with this ruthless economic democracy to offset some of the costs that are involved in change.

We can't really deal in an overall sense with the nature of

civilian technology and government investment until we have been able to make a real advance in measuring social costs. We have what we call Gross National Product but it masks all kinds of deficits. For example we build a steel mill and it is an increase in Gross National Product. At the same time we may pollute a water system and then have to have a new plant to clear up the water. The new plant is considered an addition to Gross National Product rather than an offset. We have thousands of cases of this kind. We simply don't have any mechanism which will allow us to choose rationally where we should go. We may spend fifty times as much developing atomic energy than we have for the coal industry. What would have happened if twenty years ago the government had started research on the technology of coal? Where would the Appalachian coal miner be now? Unless we have an overall allocation instrument, it seems to me that the discussion of these problems, while very intriguing, will be rather sterile.

Question: To what extent are the physical scientists aware of the social implications of their work? Is this awareness increasing?

Mr. Weinberg: We should specify whether we mean the military or the civilian social consequences. My own impression is that scientists can be divided into two groups, those who have time, or take time to concern themselves with social consequences (this group tends to be the administrators) and those who have their bellies so close to the bench that they can't or won't take time to consider broad issues.

Take the rather disastrous effect that economically competitive atomic energy may have on coal miners. I don't really think that many of the atomic energy people concern themselves with this. They are busy people, and they are not generally given to thinking in these terms. I think that the people down at Oak Ridge would become better citizens and our community a better community if our scientists became more aware of these social problems. Over the last couple of years we have made a small attempt to bridge the gap between the two cultures. Last summer we had a six-weeks seminar on science and contemporary social problems to which we invited some thirty-five social scientists and humanists, as well as Oak Ridge scientists. The seminar was a success, and I hope similar seminars will be repeated elsewhere.

TECHNOLOGY, SOCIAL CHANGE, AND THE SOCIAL SCIENCES

Question: Isn't the whole process of the development of technology on the one hand and social institutions reacting on the other a kind of Markov process which has reached a steady state? If technology gets too far ahead of social institutions like labor unions and universities, then the sources of technological change begin to decrease. The inability of the social institutions to keep pace sets a brake on the creation of technology. Social research which would enable social institutions to change more rapidly, whether it be labor unions or in the universities themselves, would in turn permit more rapid technological development.

Mr. Weinberg: Do you think there is any possibility of the social sciences progressing to the point where they can be used as a guide for changing social institutions? I think this is implied in your remark.

Comment: If the social scientist can't improve social institutions then perhaps we are stuck with this steady state condition.

Comment: I think you mustn't run the social sciences down as far as that. There is a real degree of flexibility in our social institutions. If we have the political ingenuity to use the social flexibility available, we can do a very considerable amount. For example we have millions of people in the United States who receive an income without performing productive labor, and we consider that their income is completely respectable. We have people in the Armed Forces, retired soldiers, older people on some kind of pension, millions of students who get an income so that they can go on studying. Suppose we were to introduce sabbaticals in the labor force. It would certainly help to ease the unemployment problem. If the situation is as serious as you say it is, then I am suggesting that there are available in our system flexible types of adjustment. Social scientists are quite capable of making suggestions to meet many of these problems.

THE RESPONSIBILITY OF SCIENCE FOR
INNOVATION AND ADAPTATION

Question: You seemed to assume that science has responsibility for the process of innovation and adaptation. Doesn't this push the responsibility of science out too far? Don't the initiative, innovation, and adaptation come from outside the scientific and technological world?

For example, take the complex of industry along Route 128 outside Boston. I think that it is quite possible that the initiative and the creativity to assemble these talents and apply them issued in part from the business community, the financial community, and some of the other institutions of that area. Perhaps the scientists themselves were not really the prime movers.

Mr. Weinberg: Perhaps, but I know of cases in which scientists were the prime movers. For example some people who were connected with one of the national laboratories decided a couple of years ago that it would be "fun," as they put it, to commercialize a scientific device they had developed at the laboratory. It turned out that from their "fun" sprang a vigorous company that now employs a hundred people. This caused a problem within the laboratory because the new business was based on findings paid for by the government. I should guess that if you look at the really striking successes along Route 128, in almost every case a scientist who has some of the characteristics of the entrepreneur is the prime mover. Knotty problems of conflict-of-interest are involved here. I believe some of the measures we have taken at Oak Ridge ought to help control what I call the Route 128 syndrome, without favoring any one group unfairly at government expense.

Advanced Technology and American Business: Friends or Foes?

by FREDERIC de HOFFMANN

President, General Atomic, General Dynamics Corporation

I WOULD LIKE to address myself to some of the problems and challenges which a highly scientific and technically oriented business organization encounters, particularly one having a large research laboratory as an integral component of its operation. The enterprise with which I am connected, namely General Atomic, a division of General Dynamics, was born at a time and has operated under circumstances which may be somewhat unique. This may have shaped my point of view regarding some of the problems I would like to discuss. For this reason I shall begin with a few brief observations about General Atomic in order that my later, more general, remarks and conclusions may be seen in their proper light.

General Atomic was founded as a division of the General Dynamics Corporation in 1955 and began to operate in 1956-57. It was created as a completely new business encompassing the spectrum from research to development, production, and construction in the energy field, with particular emphasis on nuclear energy. It was agreed at the outset that this business would have its own roots within its own fundamental research laboratory, so that new ideas from the frontiers of present-day knowledge could quickly carry General Atomic into other modern technological fields.

The nuclear energy field is perhaps a shining example in the civilian sector of a field that involves many scientific and technological disciplines all at one and the same time. Many may think of the nuclear energy field as primarily a physics and engineering field; but it takes only a moment's reflection to realize that it is as heavily a chemical and metallurgical field, if not more so. It was my belief in advocating the founding of something like General Atomic, a belief I hold more strongly today than I did in 1955, that it is impossible for a modern industrial production center to be good in the true sense of the word if it does not at the same time do enough research-and-development work so that there is a continuity of techniques and information all the way from the invention to the production of an item. General Dynamics turned this belief into a reality. Once you accept this belief as a basic premise, then, of course, it becomes necessary for General Atomic to have strong capabilities in many scientific technological fields in order to carry out even one of its jobs, namely, that of developing and producing novel nuclear energy systems.

Thus, we are faced with a problem of setting up an interdisciplinary team in which each discipline is strong in itself. Only then can we be sure that the individual in the interdisciplinary team will be of sufficiently high caliber. The only way I know to achieve this is to make the work within the given departments sufficiently attractive from the point of view of the discipline, and not merely from the point of view of some of the interdisciplinary efforts in which the group as a whole is engaged. After all, if one has spent many years training to be a theoretical physicist, or a chemist, or any other specialized scientific researcher, one would very much like to have the pleasure of making a discovery or advancement within that field of specialization. Thus, it becomes natural to suggest that at General Atomic there should be major projects which are, let us say, purely chemical or purely physics and which do not deal with nuclear energy. Wherever possible, of course, the departments have steered their projects in a direction that has some continuity with the main fibers which hold General Atomic together.

The main fibers running through General Atomic are several. A typical fiber relates novel energy production, energy conversion, and energy storage, and all of these are unrelated to fission energy. General Atomic, in fact, engages in: (1) the largest privately supported fusion program in the nation for the liberation of energy from heavy hydrogen; (2) the development of a new lighter weight chemical storage battery for mobile and even vehicular applications; and (3) one of this country's largest programs for the direct conversion of heat to electricity.

Another main fiber which crisscrosses General Atomic in another way deals with materials technology at high temperatures— for example such different activities as high-temperature nuclear reactors operating at 750° to 850° C, high-temperature fuel cells operating at 1000° C, and the production of high-temperature thermionic materials operating at 1800° C.

Yet another fiber crisscrossing General Atomic in different directions deals with high-energy fluid dynamics. This being, of course, the basis for the modern field of hydrogen fusion, it furthermore enters strongly in General Atomic's nuclear weapons effects work, and we encounter it again in General Atomic's Project ORION (the propulsion of space vehicles by means of nuclear explosions).

Still another fiber running through General Atomic is that of high magnetic fields. Basic to a number of areas of solid-state physics work, it is also a pertinent part of the fusion work and, perhaps more unusual at first thought, it forms the basis of a new machine tool industry which General Atomic has nursed along quite rapidly in the last few years, the MAGNEFORM development. MAGNEFORM is a good example of a field of business generated from within General Atomic itself. The intimate knowledge of high magnetic fields led General Atomic's physics department to invent a system whereby these high magnetic fields are used to generate high pressures in metals adjacent to these magnetic fields in order to form these metals. MAGNE-FORM machines are in use in a variety of industries including production lines in the automotive industry. I feel certain that this rapid development of the MAGNEFORM at General Atomic has something to do with the fact that we really practice our belief that research, development, and production should be closely integrated by the simple device of having senior scientific and technical people have a large say in all phases of the operation and not only in the traditional scientific and technical end.

We have, at General Atomic, followed a policy of *not* trying to force organization along traditional lines. A look at General Atomic's organization reveals that it is very difficult to tell what specialized field Director or Vice-President A is responsible for as opposed to that which Director or Vice-President B is responsible for. It seems important to us that our senior people particularly have not only a broad understanding of, but also a broad personal feeling for, the many different problems that a highly technical business operation encounters. In other places, particularly large organizations, this broad range of experience is often gained by rotating people through a series of assignments so that they learn the totality of the business, much as a father puts his son into the business by having him work a little while in each of many departments.

There are two drawbacks to this rotation system. First of all, for it to operate, there must already be something in existence, and at the beginning General Atomic didn't exist. Second, and more important, we wanted to, and continue to want to, avoid doing something a certain way just because somebody else had done it that way. Therefore, we adopted the plan of having our senior people oversee quite different types of activities at the same time. In particular, we wanted to assure that, if at all possible, senior people would have both functional *and* project activities report to them at the same time. This is one of the ways we believe the traditional fight between functional and projecized activities can be overcome. Both kinds of activities are necessary even within our organization where one and the same person has to reconcile the diverse activities which must coexist, and has to reconcile in himself that they must coexist harmoniously. Thus, several such people are enabled to form a core of understanding which acts as a buffer against unnecessary personnel friction or against the attempt of any single individual to drive either a functional or a projectized activity too far in either direction in order to achieve personal gain within the organization.

Briefly, General Atomic believed in increasing its staff by what I like to call the "snowball" effect. For this effect you have to have some centers for the snowballs, and I think we were lucky in the mid-1950s that there was a historical climate which enabled us to get a group of technical people together at General Atomic. Many had worked on the nuclear weapons program, either during Manhattan District days or after, and by the mid-1950s the development of both the atomic and hydrogen bombs had been accomplished. On the other hand, the development of nuclear energy for peaceful purposes was quite clearly lagging far behind. In many ways it was and is a much more complex challenge, since it involved not only making a device work but making it work cheaply enough to be of economic benefit.

Nuclear energy seemed to be a subject which ideally could be tackled most efficiently in an industry of a peculiarly modern kind. It would have to be an industrial enterprise which would allow the academic freedom of inquiry and flexibility to be linked closely and without restraint to the productive push and resources of industry. It seemed to a number of us that government facilities could not provide this atmosphere for two major reasons: 1) in the end these government facilities, in our free enterprise society, would not produce the actual nuclear reactors, and therefore the essential link between research and development on the one side and production on the other would be broken; and 2) by the very nature of governmental institutions, the requisite flexibility might not be come by as readily as it is in nongovernment run institutions.

This is then how and why General Atomic was born nearly a decade ago. Today it has a staff of over 1,800, of which more than a quarter are on the higher degree level and over a half are technical personnel of one type or another. I won't list all of the products. Just allow me to say that one of them will be at Columbia University in the very near future—a novel TRIGA research reactor—one of a large family of reactors which we have built on five continents. Our largest product in terms of size is the high-temperature graphite nuclear power system of which the first prototype power plant is now nearing completion in Peach Bottom, Pennsylvania, in a project undertaken with a very large segment of the private utility industry in the United States, nearly fifty-three companies.

And now, with this background, I shall turn to my main subject, which consists of a series of observations and a few suggestions.

SOME OBSERVATIONS

Decision Centers. I think we would all agree that it is most important that good ideas in research and development should have a chance to be heard and tried out. In the days of small projects, if the effort in a given field was a quantity X, it was unlikely to be carried out in one place but perhaps as many as ten places at $0.1X$ apiece. A technical person operating within one of these units had a strong chance to have good ideas not only listened to but also tried out and carried to successful conclusions. What has happened over the years? Not only has the quantity X grown, which is what we usually focus on, but, equally important, X gradually became subdivided into fewer and fewer units. The ten units became five, and the five units became three, and so forth. As the subdivision of X tends to be eliminated, it is harder for individuals to have bright ideas tested at an early stage.

Now, let us look at the effect the increasing time-duration of projects has upon research and development. Times have indeed changed. No longer does a "crash" three-to-five year program seem a paradox to us—yet three to five years is a long time when measured against the span of truly effective working years the average scientist and engineer have. Be that as it may, the projects do take that long. The danger that such long-term projects bring with them is that it tends to become difficult to "change courses in mid-stream"—once again an effect similar to the first we examined that tends to operate against new ideas.

As new projects have grown larger and lengthier, there has indeed been much fruitful government-industry cooperation in technology. In the balanced role of this government-industry relationship lies a mechanism which can do much to retain the diversity of effort achieved in a more natural way in the pre-1940 era of small projects. Basically, now that projects are so large and costly, both industry and government face the same question: "Have we picked the right project?" There has been a tendency to answer this question by trying to find ever better methods of making such decisions for the country as a whole. I believe that this alone is not enough.

I think we all agree that no one set of people, even those of our friends who have the difficult task of carrying on some of that central coordinating job for the country, is wise enough to forecast all of the most fruitful avenues to be followed in research and development. Therefore, I think it is most important that there be several decision-making centers regarding large-scale research-and-development work rather than only one central one, regardless of how excellent a job the central one may do. With the projects as large as they are now, only one segment of the country besides the government can provide such centers of decision—industry. Thus, I believe that we in industry have a unique duty to help constructively to assure good research and development.

In research and development it is often not a "mistake" to follow the currently unpopular line of research because research and development consists of fluctuations and not a straight-line path to success. One must recognize that the government, of necessity, must operate under more severe restrictions than industry, because the limelight in which the public official operates makes it more difficult for him to change course.

Then too, take the difference between government and industry caused by the nature of the four-year governmental cycle. This cycle had little or no effect when long-range projects had an average duration of six months so that several could be tried in succession and there was a good chance that one of them would be successful within the four-year span. As projects stretch to several years' duration, this clearly becomes more difficult. It is perfectly natural and human that under these conditions even people striving for the best solutions become more cautious and less inclined to be deflected by "new ideas," because they serve in the government "limelight."

I believe that industry inherently does not have to operate under the same time pressures as government because time and time again the long-range interest of industry has proven to be best served where consistent long-range technical developments were undertaken by that industry. Thus, I believe it is the duty of industry, be it in the manufacturing or distribution end of it, to set aside some part of its funds for long-range research. (I shall return to a suggestion as to how this might be encouraged in practice.) Only in this way can we be sure that there are enough decision-making centers in research and development throughout the country.

The Fallacy of Compartmentalization. Note that in discussing decision centers I have purposely avoided the fallacy increasingly accepted in this country to try and divide research, development, and production into neat, separate compartments and then to go one step further and talk about the fact that research can best be done in one atmosphere, development in another, and production in yet a third, with the ever recurring suggestion that research had best be done only in government laboratories and nonprofit institutions. This is what I like to call an "attractive fallacy" and has the same fascination as the "attractive nuisance" of an unguarded swimming pool to a three-year-old child who happily falls into it and drowns.

Let me state again that I believe truly efficient production of technologically advanced products needs closely linked research and development in the same organization. Furthermore, one should not have preconceived notions where research can best be carried out. The necessary immediate environment—yes, that must be right. But the sponsorship of that work does not guarantee good people. By and large, funds should always follow good people, and we must guard against the thesis which is gradually evolving in the public mind that public funds should not be used to support basic research in industry whose main function in the popular view is to bend hardware or at best to do development. I feel strongly that research, development, and production of an advanced product are closely linked, and that these functions cannot be split apart in an arbitrary manner.

The trend towards compartmentalization is, of course, fed by the further popular fallacy which asserts that the "pure" scientist has a great disdain for the "applied." This is rarely so among the truly good scientists. What the truly good scientist seeks is merely the assurance that all problems, as far as possible, are approached in a scientific manner, and that he personally may work in an atmosphere conducive to creative work. He may not choose to work on certain problems himself, because he believes he will be more creative in another field, but certainly this should not be interpreted as disdain for the particular problems in question. More often than not, the problem actually does attract him, but he has both a quite proper disdain for the methods used to find solutions in the organization he is asked to work for and a perfectly healthy personal prejudice against using such methods. It is the job of any good scientific and technical management to understand the latter point and to find ways of properly approaching applied problems so that truly scientific and creative people can be proud to be associated with the effort and be interested in and get satisfaction from carrying out the work.

Management and Decision-Making in Industry. I believe that there are a number of things in "traditional" American industrial management which make it quite difficult for industry to take full advantage of modern technology and which in some cases even result in its being misled into believing that it is very "scientific" in its approach. Let me mention a few.

Some of American industry have adopted the trappings of the scientific method without understanding the very essence of science. Probably the most important single realization of modern science over the last two hundred years is that it has tended away from its earlier mechanistic tendency to find "*the* solution" and the belief at times that all that was to be known was at hand. As late as the turn of the century, a leading physicist believed that physics had become merely a question of making measurements more precisely because nothing else was left to be discovered in that field. Today everybody truly involved in science realizes that change is the most vital part of the scientific progress. In effect, you make a theory good enough to explain the situation as it is, and just as soon as you have it you try and probe for places where the theory breaks down in order to make a new theory that better fits all the facts, including all the newly discovered ones. American business by and large certainly does not recognize this. It conducts an eternal search for finding perfect stereotype solutions to given problems and is very unhappy when new facts arise that demand a change in its little world.

Strangely enough, some American businesses believe that they are very scientific because they have begun to apply one element of the scientific method, namely, the use of statistics. Even in this respect, I believe that the scientific method is badly misapplied. Valid statistical generalizations are not created by taking a few isolated facts and rehashing them in so many ways that they finally give the appearance of many facts which can be statistically analyzed.

Closely related to the question of not recognizing new facts when they arise is the impatience of some present-day American business leadership. Most business leadership in this country is still nontechnical. Therefore, even though the times have changed drastically and the technical complexity of projects now demands that longer periods of time be consumed, business managements tend to think in terms of their own personal time spans and without regard to the true, natural time scales involved. In the long run, this obviously must and does lead to faulty decisions by some industries, because of impatience and the resultant waste of effort.

I know of no cure except to get technical people into positions of responsibility. In few fields is it helpful if the man making the decisions does not intimately understand the underlying facts. In highly technical fields which ask questions of nature, knowing how to ask questions of nature and get valid answers are the underlying requirements. Note that I am not suggesting that the decisions be made without an understanding of other nontechnical facts such as regulatory, legal, fiscal, and other problems. To the contrary, anyone making the overall decisions must be able to understand these as well. I am merely objecting to the fact that it is quite fashionable to leave out one huge factual chunk, namely, the technical one, because "I didn't have a good math teacher in the third grade so I couldn't be expected to learn or understand facts based on science."

I believe that in this respect there is unfortunately considerable difference between highly technical enterprises in America and Europe. By and large the highly technological industrial companies, in Germany for instance, do place technologically highly competent people in their senior management positions. At the very minimum, they openly respect high-level technical competence. In the long run, companies such as these will show a decided advantage over some of their American counterparts in the free world competition once the long after-effects of the last war have lessened. This will be particularly noticeable once technological superiority places countries on a more even footing, lessening the advantage of more ample natural resources which, coupled with the free enterprise system, has given America an advantage for so many years.

European versus American Industrial Research. It is difficult to generalize about such a broad topic, but here are a few observations somewhat at random.

There is no such thing as *the* European approach to industrial research. It varies quite widely. Perhaps the most striking common characteristic, however, is that European industry in general is more reluctant to spend money on research which is not immediately product-oriented, i.e., they do not as a whole do a great deal of basic research in industry. This probably has its roots in two quite diverse situations: the first is that European industry until recent times did not have so large a single market to serve without boundary difficulties as did American industry, and therefore the individual industrial unit was considerably smaller by comparison and felt that it could not gamble much of its income on research; the second is that European industry and European universities have been much more closely related than have American industry and universities so that the European businessman felt he had university knowledge readily available to his enterprise. Whether the latter situation was generally true or not I don't really know. There are, of course, some outstanding examples of such close relationships, namely in Germany.

Let me emphasize that there are now some European companies which are an exception to the rule I mentioned. For instance, a company like Siemens in Germany does have very large laboratories and does concern itself with research in many areas in a fairly fundamental way. However, among the electrical manufacturers in Europe, it is difficult to find another company that matches the Siemens broad research capability.

I have already mentioned that in Germany there is perhaps a closer link between universities and industrial enterprises than in other European countries. There are several ways in which this shows up. To begin with, there are more technically trained people at the top of companies in Germany than elsewhere, something I have already alluded to. Secondly, I think it is important to recognize that, in addition to the university departments in a given science, there are the so-called Max Planck Institutes in a variety of scientific disciplines. These are not supported by the universities but rather by the Max Planck Society for Research, something that goes back to the original Kaiser Wilhelm Society. In the Max Planck Society German industrial leaders have a large voice in the affairs of the Society although much of the funding comes from the state, and the scientific programs are administered quite independently by the institutions. In addition, German industry through a "donor's association" supports pure research by making grants to various research institutions including the Max Planck Society. I know many a German industrialist who is proud to be a member of some committee of the donor's association or of the Society.

I believe this shows an exceedingly important facet of German industrial development, namely, the simple fact that German industrial leaders have a deep-seated respect for knowledge and the search for knowledge. I wish I could say that about all of my American colleagues in industry. Of course, one must not make sweeping generalizations. The big American foundations which have been created by American industrial leaders, of course, have supported much research in the United States and abroad on a large scale and in an independent manner. On the other hand, I still believe that there are large segments of our own leadership in industry who still don't show deep-seated respect for knowledge or the search for knowledge. In those cases, there is still a lot of the "rah, rah" spirit left which, when you dig just a little bit below the surface, really disdains scholarship and the search for knowledge. I need not point out that the respect for scholarship and the ability to engage in scholarship are not one and the same thing. This I believe is something which has been well recognized in Germany, and in Europe for that matter, for centuries, so that scholarship and other pursuits can live together harmoniously. In America it seems that the best defense for the man who is not suited temperamentally—either by virtue of the way his mind functions or simply by deliberate choice—for scholarship is to belittle that scholarship. I submit this is an uneasy truce between advanced development and some of our industrial leadership. I leave it to the sociologists to understand the deeper meaning of this problem, which undoubtedly has to do with the fact that our country is a relatively young one. Young as it may be, however, nobody will deny that the United States now plays one of the dominant roles of the world and with that role surely comes responsibility. The responsibility of closing the gap rapidly is an essential one.

The patterns of industrial research in England and France are difficult to untangle because in some fields a good deal gets done by the government itself through its inhouse organizations. This is particularly true in France where, for instance, in atomic energy there have been numerous discussions between the French government and French industry. Also in France research in the consumer field, such as the gas industry and a number of other fields, is carried out by government organizations.

In observing European industrial research and its relationship to European governments and trying to understand the pattern, one must realize that it is not possible to understand it simply in American terms. Here in America we have a homogeneous country with various segments such as industry, government, and the university carving out certain segments and fields for themselves. On the other hand, Europeans have tried hard to build a more integrated Europe through the Organization for Economic Cooperation and Development, the Common Market and its sub-organizations like EURATOM, etc. Thus, in Europe all the normal processes of development of advanced technology always have in them a flavor of politics in the truest sense of the word. Obviously, the tug of war between different states in America to get a certain scientific or industrial research installation put within a particular state is only a small sample of what happens when Italy, France, and Germany argue about where a new industrial research facility ought to be placed.

To get a feeling for the pattern of governmental research in Europe I refer to a work by Pierre Piganiol and Louis Villecourt, *Toward Scientific Politics*, as well as the work of Arnold Kramish of the RAND Corporation. Kramish has compiled some very interesting statistics on the subject of European research both governmental and industrial. In particular, I would like to cite four conclusions that Kramish has drawn from his recent study. They are as follows: 1) The European research effort on a per capita basis is less than one-half that of the United States or the USSR; 2) The costs of supporting research and development, per qualified worker, are rising sharply in every country; 3) The rate of increase of research and development expenditures in every nation of the EEC is rising at several times the rate of growth of the G.N.P., which is typical of the situation in the United States, United Kingdom, and the USSR; 4) Industrial research and development in Europe is probably increasing at a rate higher than industrial research and development is increasing in the United States.

Note, however, that the last conclusion, of course, is about the rate of increase rather than the absolute amount. Since we have already seen that the absolute effort in Europe is less, particularly so in basic research, it is perhaps not surprising that equalization between Europe and the United States is taking place in this field as it is in many others.

In many European companies the family corporation still maintains itself even in the very largest enterprises. In America we tend to regard this as patriarchal and almost by definition as not a very good thing. There are two sides to this question. Certainly the advantage of the large American corporation is its broad ownership. However, the advantage of the European family corporation, or at least family identified enterprise, is that quite often it builds a peculiarly strong moral fiber into the top of the business. After all, the family knows that when the business does something wrong, people will not regard the corporation as the "the corporation" but associate it with some wrong-doing on the family's part. This way, even a large corporation cannot become "apersonal." This way, too, the management often feels a stronger personal type responsibility even to a large work force than the American corporation for whom there is a natural tendency to feel the push of the anonymous stockholder to get a large immediate return at the expense of long-range or human considerations.

Something-for-Nothing Philosophy. I believe that the point at which American industry and American advanced technology become unwitting adversaries is the point at which there occurs what I like to describe as "gold-at-the-end-of-the-rainbow fever." Maybe I can explain it this way. In the 1950s science and research became such a vogue in American corporations that there was an underlying and never-expressed feeling that this particular activity could suddenly lead to "quick profits." Perhaps this feeling harks back to the fact that most managements are not technical and therefore like to maintain the illusion of the sequence of the "sudden invention" followed by the "sudden profit." I think this mode of thinking shows itself over and over again in the tremendous search for "fall out" products from other research. Such "fall out" products do indeed exist, but they are no better and no worse than carefully conceived research products undertaken for their own sake. Why then do they have such tremendous appeal? I think it is because they have the "something-for-nothing" element which we in America love so dearly. Is it because some types of selling techniques of the last thirty years have crept into our society so thoroughly that the 99¢ bargain is looked upon as so much better, of and in itself, than the fixed price tag for proper value paid for proper value received?

This problem in American industrial research and development management has extensive ramifications. It probably does not allow society the full utilization of inventions already made. There is a really vicious circle at play here. Suppose a new invention is created or at least a new application of a fairly basic principle is made. Few American companies have the inclination to explore the value of such an invention with patience and depth. Note that I did not say with lots of money, because these are not always synonymous. In fact, it is often not the amount of money that causes the trouble. It may be that one million dollars is needed to explore many of the facets of a given invention. Now, I submit that it is often much easier to get that one million dollars spent on a one-shot "gold-at-the-end-of-the-rainbow" chance that the big market and payoff will be hit which will make everybody happy than it is to take that same million dollars and expend it over a period of five years to explore many facets and perhaps end up with two major products at the end of five years. The impatience factor is really enormous. As a result, it is often impossible to get important new technological inventions off to a good start. Previous speakers have already made this point. I wish to underscore it because the circle can become so vicious that the product development is never started even though the given industry and society would benefit.

The gold-at-the-end-of-the-rainbow philosophy has some humorous ramifications. This is best known to many in the example of nontechnical people or insufficiently skilled technical people in some companies that continuously come to the technical people with new ideas which are sure to make a million overnight. It invariably turns out that these ideas have originated from some third-order or even crackpot scientists and technologists known to these people who then take the time of serious research-and-development organizations to prove, painstakingly and politely, the item in question to be nonsense.

The Problem of Stability. The preoccupation with the gold at the end of the rainbow leads to another deep-seated difficulty in doing good research and development within the framework of American industry. It is not amiss nor a false privilege to state that good research-and-development people need a tranquillity of mind to concentrate on the problems they are trying to solve. Many people have observed that the university gives the research worker this tranquillity and sense of continuity. The more the "rainbow" philosophy takes hold of industry, and it could well be on the increase as a large segment of industry eventually shifts from defense to civilian production, the more the individual research-and-development worker finally gets to feel this lack and the greater the intrusion on his tranquillity and consequently on his ability to perform the very job he is asked to do.

Great restraint is necessary lest industrial management examine and reexamine the results of research too frequently and too continuously. The uncertainty principle does apply to this situation. The mere observation does disturb the situation and changes the resulting measurement. Appraising and steering research must be a very delicate operation lest one destroy the very tranquillity so necessary for certain portions of it.

The problem of tranquillity in the scientific effort probably goes even deeper. There is often an unconscious (and I am sorry to say sometimes even a conscious) effort on the part of a leader in a group effort to assert his leadership by making it clear that he can impose his will on those working in the group. This surely is the antitheses of making productive and creative individuals act productively and creatively. I think it is impossible for an industry to take a set of creative people and hope that one can both "get the boys in line" and, after that has been done, have them act creatively. I submit that this is too fine a line to walk. I am sure that creativity needs a certain continuity in time and that the very fact that it is destroyed for a certain length of time probably makes it very difficult for the individual ever to be truly creative again. Let me put it another way: creativity and organization for the sake of bringing out group efforts can be harmonized; creativity and regimentation cannot.

SOME SUGGESTIONS

I will now offer a few suggestions for consideration. For the sake of simplicity, I have put them in the form of recommendations. Of course, they need a good deal of thought before actual implementation.

Research and Development "Building Blocks." I believe a radical change must be made in the way development work is approached. Let me explain this by going back to the "success" of the scientific method. To many, the essence of the scientific method is simply that it enables one to predict certain results from certain facts. For instance, if I have Fact A, I can predict Fact B. *However,* it is important to note that this theory may have been developed by *experimentally* testing whether Fact C leads to Result D, not by taking Fact A and seeing whether they experimentally lead to Result B. The latter process is really simulation. Simulation is a narrow and limited process because almost by definition it requires precisely what it says, namely, to simulate each situation.

When science was quite "simple" its tools were simple, and so even when one searched for simple facts without complex tools it was an arduous process. Therefore, science tended away from "simulation" towards true "theories" where B followed from A because Fact C leading to Result D could be explained by the same theory. Modern large-scale development, I believe development—although done under the cloak of the scientific method—still involves far too much simulation. Let me illustrate this in reactor technology. I wish to build a complex power reactor costing tens of millions of dollars. It is natural that I should build a zero power critical assembly for it as a first step. Now, if I merely build a critical assembly which is a small-scale simulation, I may or may not be lucky in extrapolating it correctly toward the big power reactor. After all, if I think of my facts as a series of hills and valleys, I may just by chance sit on a very high and narrow peak in my simulation experiment (in this case, the critical assembly) and the slightest movement away from it lands me quite promptly in the valley. After all, how do I know if I sat on a ragged peak or a wide plateau if I simply had a simulation experiment. By the same token, even if I succeed in my simulation experiment, my results most likely only helped in developing *one* big power reactor. For another slightly different power reactor I would have to go through the same steps all over again.

Suppose, however, I apply the scientific method properly to the critical assembly. In that case, I will try to make quite different experiments to understand the results of various experiments that I can perform on the critical assembly, which are properly quite different from conditions in the real power reactor. If I can find theories to explain the results of these experiments on the critical assembly, then I will truly be able to say that I understand the critical assembly. At that time only will I have formed a new building block comparable to the building blocks of science, but this time it will have been a building block in development.

I firmly believe that we in this country could save hundreds of millions of dollars, perhaps billions, in our development effort if we learned with patience to create "building blocks of development" and not to step beyond these until we had assembled enough in a given field. This would then be analogous to the way science operates. Science doesn't draw general theories from one or two facts—it takes a whole series of facts each of which is *well* understood and tries to correlate these facts to a theory. I submit that what we really need is to have enough similar building blocks *understood* in the field of development. However, it is equally important that industry understands this. Otherwise, industry, too, will continue to operate on luck rather than on solid footing in the development field. In the end, this leads to impossible costs for industry and the nation as a whole.

Responsibility of the Scientist and Engineer in Industry. I have spoken of the continued need for decision-making centers for research and development in industry. However, I firmly believe that these can be truly effective for the nation only if industry as well as government is motivated by one and the same desire, namely, that the best research-and-development work be done for the nation as a whole. I believe we in industry have the responsibility to see that there is another set of funds, other than that of the government applied in research and development— but the motivation for its use must be the same. This is easier said than done.

Are there means which will ensure that industry will follow a logical course in research and development? I believe there are. It is here that those who happen to be in industry and whose backgrounds are in science and engineering come in. I think they have to adopt a certain code of responsibilities and I would like to suggest a few.

1) Recommend only projects whose technical merits you truly believe in.

2) Do not pre-plan work so completely that there is no room for innovation.

3) Don't budget research and development without contingencies.

4) Terminate certain research-and-development projects. This is too easily said and very difficult to carry out. The reason it is difficult is that usually termination comes about merely as a budgetary pressure because money is being fully budgeted and spent, and there is little chance to try out new ideas. Actually, to my mind, the best way to terminate projects is by having deliberate flexibility in budgets so that enough new ideas can be tried out to make it obvious that a new project has a better chance of getting to a goal than did the old one.

5) Don't forget that research-and-development projects depend on people. In a university the creative person shapes work not only in his own field, but, by definition, shapes the university as a whole. In government service, the creative person shapes work not only in his specialty, but in the policies of the government. Similarly, industry will create the best atmosphere to attract creative technical talent if this talent has a voice in setting industry's goals.

6) Encourage industry to open up truly new fields. It is important to remember that, contrary to the popular or sometimes even engineering view, it is often *easier* to make a "quantum jump" then to edge up on technology. This point of view is easier for universities and government to accept than it is for industry as a whole. After all, much of industry has traditionally simply carried out production by existing means. Therefore, it isn't a pleasant prospect to throw away today's productive tools. Nevertheless, this is a situation industry must face.

A New Tax Treatment. I have pointed out the strong trend towards having fewer and fewer decision centers for research and development projects, and I would like to address myself to one possible way to reverse this trend. At present there is, of course, a somewhat vicious circle at play: Most of the research-and-development funds are spent by the government, and the tendency is towards less duplication in the government and therefore fewer decision centers controlling the ways in which this money should be spent. The industrial firms help supply this tax money which is used to support federal research-and-development programs. As the clamor for more research and development increases, the federal government must collect more taxes to finance it. This in turn reduces the ability of industry to do research and development itself and thereby provide additional decision centers. How can we reverse this trend?

I believe the tax situation itself could be used in dealing with this problem. In most of the tax situations it is very difficult to pair off equal amounts in receipts and expenditures and thereby not to unbalance the federal budget further. However, in this particular case, I think it would be possible to take a small amount of the existing federal expenditures in research and development each year (say half a billion dollars to begin with), and offer American industry an equivalent tax reduction provided that amount were directly spent on additional research and development. I know such a plan is not easy to work out in detail. I know that cries of "too easy cheating" under such a plan will arise. But I think it is worthwhile to devise such a plan with all its attendant difficulties because it is one of the few ways I know that will directly stimulate more research and development associated with our actual production centers, and at the same time will give us the necessary new and additional decision centers.

I know this idea is radical, but only because we have been told it is necessary to control research-and-development funds more closely and from a more centralized source. I again submit that in the long run we will gain far more from extra centers of decision-making and from duplication than we will from a mechanical, federalized, planned research-and-development effort in a country as large as the United States. The latter just doesn't allow any room for human failings, and the addition of decision centers allows the law of large numbers to play a part in smoothing out these human failings.

Logic as a Necessary Link. Finally I would like to make a very obvious suggestion which perhaps is still worthy of a brief discussion. Almost all the things which tend to drive advanced development and American industry away from each other rather than toward each other, I believe, go back to the fact that individuals in society have been brought up in different compartments and with different educations.

Although many of my audience may be educators, I would like to make a suggestion for the American educational system. The American educational system must try to find a common denominator for many individuals which does not involve cooking and sewing in high school and football and rowing in college. I suggest that perhaps this field should be the field of logic and its outgrowth, the scientific method. These abstract ideas should be taught to one and all as a means of giving them a common heritage and a way to communicate with each other in a language that allows for differences of opinion but not for misstatements of fact so that the cleavage gets wider. I believe if this were done those of society who, for instance, would later on specialize and excel in the liberal arts, would find it very easy to understand the true meaning of science rather than thinking of scientists as engineers with slide-rules atop a test tower who have no appreciation of art or literature.

If this logical training were accepted as a common denominator, I believe the "seat of the pants" way of running enterprises, so highly regarded by some, would soon disappear, and a good thing, too. Seat-of-the-pants decisions, I believe, are a luxury which one can afford only when one has lots of pants.

Let me close by saying that I personally have such a deep-seated respect for education, knowledge, and the search for knowledge that I am sure a way will be found by which advanced technology and American business will become close friends and that a generation hence the title of my paper will be an enigma.

DISCUSSION

THE LINKAGE BETWEEN RESEARCH, DEVELOPMENT, AND PRODUCTION

Question: I would question your linkage of research, development, and production. Are you not generalizing from a particular case, an organization that engages in a particular kind of scientific work?

It seems to me that in many aspects of research a division of labor might be desirable. I can see where scientific progress might be retarded if scientists had to worry about production, sales, and other elements.

Mr. de Hoffmann: I tried to say that this linkage applied only to advanced technological products. But as to your main point, I have a really different view. I think there is intellectual stimulation in combining these activities. Sometimes, I think it's quite interesting to have problems overlapping with other fields. For example, take mathematics. When the Egyptians solved mathematical problems, they solved them because they wanted to know something about practical things, but this didn't deter them from doing exceedingly good pure mathematics. In the same way, I think somebody who works on a metallurgical problem very often is stimulated by knowing that after he has solved his own problem there is still the further problem of determining what is to be done with his solution. Research is such a very broad thing. There is really no such thing as applied research and pure research. There is only good research and bad research. Once you take that point of view, you can get a certain pleasure in working on things having practical application. Most of us occupied with pure science, whatever we say, still enjoy having our work applied at the end. Moreover, it does sometimes lead to better research to know where you're going.

In highly advanced technological industries, it is wrong, I believe, to separate development work from production. An example is nuclear fuels. Our company is a major producer of these fuels. Initially some persons in our organization felt that research and development had to be kept separate from production. However, when production problems began to arise, we were fortunately able to bring our research people back into the production line. Our research people actually enjoyed helping put the factory on its feet. In a traditional organization, it would have taken at least four months and a stack of memos before the research and production people would have even talked to each other. In this organization it took only a few phone calls.

Comment: I think it's true also in the social sciences that the theory gets better when social scientists are forced to engage in practical problems.

Question: Do you think the universities should engage in development work as well as research?

Mr. de Hoffmann: I believe that it is "people" who should engage in particular work and not "organizations." Thus, to answer your question, if there are people at universities who are particularly adapted to doing development work, by all means let them engage in development work. More generally, industry should engage in pure research and universities should engage in development only if they have the proper people to direct and do this work.

ORGANIZATION AND STAFF IN A COMPANY ENGAGED IN AD-
VANCED TECHNOLOGY

Question: You mentioned that in the course of your normal day, you talked to people who have very sensitive technical positions, but who nevertheless do not have organizational responsibilities. Does this reflect some solution you've found which allows a technically trained person to rise in your organization without loading him down with administrative functions if he doesn't want them?

Mr. de Hoffmann: One solution is to be sure that a senior technical person can go up the ladder salarywise truly independent of his administrative responsibilities. But there is one problem we certainly have not solved. That is the problem of how to make him a professor or the equivalent in terms of prestige. This is a difficult problem for an organization that tries to recognize ability, yet doesn't want to do it just by salary alone. Certainly, we haven't solved it. We call some of our most senior people who do not engage in administrative work and who do not carry heavy administrative duties "senior research advisers." But, that's just not the same as being a professor. I don't know any way out of this problem. It gets solved, in part, by the fact that some of these senior people at General Atomic do get offered positions of professional rank at a university. Fortunately for us, a number of times now a man turned down such university offers, and the mere act of rejecting the university offer gives a man the sense that he has reached this level. The other way we try to solve this problem is to permit a man to engage in whatever research interests him, which, I guess, corresponds to his being a "university professor," or, as some call it, a "professor-at-large."

Question: Having men in the same organization who represent diametrically opposed interests would seem to have obvious advantages, but where do you find these people? What general backgrounds do they seem to have?

Mr. de Hoffmann: The most adaptable people turn out to be some of our scientific and engineering colleagues. Of all the other people involved, the lawyers and broad fiscal people are the most adaptable. Sometimes, businessmen are the most inflexible. This, I think, reflects the different types of training which each group receives.

Question: Do you have to develop people of broader talents, or do you find them ready and available?

Mr. de Hoffmann: We have to develop them, and they sometimes dislike it at first.

Question: What kind of cost does this involve for your company?

Mr. de Hoffmann: I can't measure it for you, though I'm sure it is significant. But it has paid off for us in the sense that it maintains respect among the various members of our organization from one end of the spectrum to the other so that we do not constantly have to deal with personnel problems.

I think we can act faster because, for example, if the lawyer really understands all the technical material, or if one of our technical people understands the legal problem, they each go out themselves and do something about it instead of waiting for the other to act.

Question: Given the ideal structure of the organization that you posit, does it assume an unlimited source of highly skilled manpower? Secondly, what sort of competition for this kind of talent have you met?

Mr. de Hoffmann: The system I posit does strongly believe that there are certain people that can do certain things, but not everybody can do everything. The second thing I would say is that I believe that there is much more latent creativity in people than is generally admitted. We constantly try to upgrade our people.

The competition for personnel is not as troublesome as it might be. We do not try to "buy" people by salaries. We have tried to attract our people by providing this combination of industry and the academic world.

RESEARCH DECISIONS AND DECISION CENTERS

Question: You spoke convincingly of the desirability of increasing the number of decision centers, and you also spoke of good research and bad research. How can we get correct decisions about research?

Mr. de Hoffmann: I would like to have the natural process of selection take over. What I want to see is industries take entire fields of research, try various techniques, find out that some of them lead to absolutely nothing, close these down, find that other things do bring success, but be willing to give them some time, because these methods may take long to develop. I don't want to force this process.

Question: How, within a good organization where it is necessary to be discriminating about people and projects, do you make good decisions?

Mr. de Hoffmann: I think it is possible to get good decisions only when the people at the top have a broad knowledge of the field that the company is engaged in. I can only use the analogy of the university. There's a fairly well-established way to make decisions in a university. A man first works on a particular subject, eventually becoming a professor. This gives him the privilege of making decisions precisely because he knows what research has gone on in his area. However, this system of natural selection does not operate very often in industrial research and development. As you can see, I favor a system whereby those who have done active work in a particular area are given the responsibility for making decisions. I do not know any other way of selecting people to make these decisions. I am pessimistic about the possibility of training professionally scientific research and development managers, because I believe they would lack this ingredient of knowledge and experience. I think they would just not have the patience, they would not understand the errors that would invariably occur.

Question: It is easy for people to carry on unpopular research in a world which is atomized, where everybody can move out on his own. But where large organizations and large sums of money are involved, how would you insure that unpopular research can be carried on?

Mr. de Hoffmann: Only by getting more decision centers. As soon as the decision-making process is centralized and a man has to allocate two or three hundred million dollars at a time, he cannot afford to support unpopular research.

Comment: One of the threads that ran through your talk was the need to have people with technical training or people who appreciated science and technology in higher positions in American industry. It's not quite clear to me how much of an improvement would be made just by changing the quality of personnel.

Mr. de Hoffmann: I'm worried about the practice of perturbing the results of research by questioning the researcher. Unless one has experienced this directly, it is difficult to appreciate how frustrating this can be. I don't see how somebody who is trained only as a research manager, who has been told that "tranquillity" is a problem but who has never personally experienced it, can know what it is like to be just at the edge of finding the solution to a problem and have the phone ring with the director of research asking him whether he has solved it.

RESEARCH AND DEVELOPMENT BUILDING BLOCKS

Question: Is there any lesson that can be learned from the kind of "big science" effort we have gone through already, for instance, the efforts to get controlled thermonuclear reactions or nuclear powered aircraft? Are there lessons we can learn about how to manage technical enterprises?

Mr. de Hoffmann: The things I've said apply to precisely the cases that you have mentioned. I was not surprised at the failure of the nuclear-powered airplane. I don't understand the kind of reasoning that says that if one works long enough and spends enough money on a nuclear airplane, it has to work. This was an example of attacking a problem by "simulation," but to my mind such things cannot be attacked in this way. The necessary building blocks are often missing! This is a lesson which must be learned, and which, I think, may have been learned from the nuclear airplane program.

Question: Do you think this is a problem of personal competence, rather than a problem arising from the institutional structure?

Mr. de Hoffmann: I think it is the problem of relying upon simulation. It stems from the deep-rooted belief of some old-line American industries which has spilled over into American government, that if enough people are working on enough things, problems will inevitably be solved.

Question: Is the simulation problem basically an educational problem or an institutional problem?

Mr. de Hoffmann: It has to do with the question of respect for knowledge. Simulation is a kind of "rah, rah" technique where it is believed that if only enough people are put on to the problem the problem will be solved. If you have respect for knowledge you instinctively dislike this method of approach. I think the way to bring about some of the necessary changes is to get people who have broad knowledge, who have an appreciation for scientific method, and who, therefore, will form new structures that do not kill the existing structure, but add to it.

Question: How does basic theory get translated into a finished product?

Mr. de Hoffmann: Your question assumes that first of all there's a vast set of basic knowledge. Then a company comes along and decides to work on something, using part of this knowledge. My view is that often a company does not have this kind of knowledge. There's another way of approaching problems. This is to recognize a need for something, for instance energy. Our company is a specific example. General Dynamics, to diversify, had to take on a very large new field. In the energy field the demand for electricity doubles every ten years. We started out with this need and then created an organization which tries in the most scientific manner we know to attack that problem.

I don't believe in the attitude which says: "Here's a lot of scientific fact. Now I can just take eight of these facts and make a product out of it." The difficulty arises because you may not have three other facts that are needed. What you need are research-and-development building blocks. More products and more research and more development will be brought about, not by expanding the scientific base, but by developing research-and-development building blocks. But you have to have some end objectives to do this.

Question: In what sense then does your company engage in basic research?

Mr. de Hoffmann: In two senses. First it attacks the problem it is engaged in solving. For example, it tries to understand what goes on in the scattering of neutrons in graphite and beryllium. This is exactly the same kind of work that might be carried out at Columbia. The difference is that we do it with a specific purpose in mind, but the technique and the type of scientists involved ought to be very much the same and hopefully just as good. Secondly, we have to create projects which fall more into the fields of specialization of our personnel. In this sense we do basic research like a university. We work for instance on the age of the universe by working on the problem of what distribution of elements is found in meteorites. How does this fit into our picture? First, it keeps very good chemists around; secondly, it uses a technique which we happen to be very interested in, activation analysis, in which we do some commercial business also.

CREATING ADDITIONAL RESEARCH CENTERS

Comment: You have pointed out the extremely narrow research areas within which very large companies can find it profitable to move. For example, General Dynamics decided that energy was one of the few fields which they thought they might find profitable some day in the future. I think that this is one of the major constraints in American industrial research-and-development work. For example, large chemical companies, on the whole, are frequently uninterested in pursuing certain fields, because when they make their forward calculations they just cannot see how the field is going to pay off, given their market positions. They would have to be enormously impressed by the possibility of a transformation of the market before they would be willing to enter a new field.

Question: Do you want to make structural changes to help private companies make these big investments in research?

Mr. de Hoffmann: I want a conscious decision on the part of our society that it is necessary to have more research done in the private sector in order to get more decision centers. We have gone so far as to make a tax cut to stimulate the economy. Now I say, the time has come to make a tax cut to stimulate more decision-making centers.

Question: It seems to have been accepted as a premise of this discussion that major breakthroughs, major research, major development must be accomplished by large institutions that have the resources to carry out research and development. Yet most of the really new ideas have been born in small companies, extremely small companies, which lack the resources to develop them, and were then acquired by a large company. Is there something that can be done to provide greater resources and incentive to the small company to undertake basic research?

Mr. de Hoffmann: In answer to your question I might say something about the patent situation. Proposals to reform the patent system so that private companies under government contract would not be allowed any domestic or foreign patent rights have become very fashionable these days, but they would play directly into the hands of big companies. Actually, the man this would discriminate against is the inventor in the basement. Such a man, if he had real patent protection, if he had a bright idea, might become an entrepreneur and start a new firm or industry. But today, if he works for certain agencies of the government, his idea is public property and is available to any big company to exploit. This hits hardest at the little company.

Question: I'm struck by the degree to which the image of science that emerges from what you have to say is different from the conventional image. If this difference is as great as it seems to be, can anything usefully be done to create a different public conception of science? You suggested, for instance, a change in tax laws, but this would require public understanding of why this should be done.

Mr. de Hoffmann: You probably would have to change the industrial image first, I suspect. But I think this would happen once you had more industrial management eager to engage in more research activity because of its long-range rewards to that industry.

Science and the Civilian Technology

by J. HERBERT HOLLOMON
Assistant Secretary of Commerce for Science and Technology

I WANT TO REVIEW some of the developments taking place in our society as they affect technology and national needs. First, as society becomes increasingly more complex, its needs become increasingly more social than individual or private. More than half of the new jobs created in the last decade have resulted, directly or indirectly, from governmental rather than private effort. This is because the activities which have to do with education, medicine, or defense and space, including those carried out by private companies, are generally initiated and sponsored by government. Many organizations dealing with military and space problems, for example, are in no sense private, since their only potential customer is the government. This is a new phenomenon, where so much of the new employment is created to meet complex social needs rather than private wants.

The second development is the large number of our people who are now urban. Three-quarters of our population live in cities or in areas directly associated with cities, and their problems are the problems of a complex, urban society.

The third element is that the majority of our workers are now engaged in service industries. More than half are devoting their efforts to health, education, recreation, government, repair, and distribution—to services rather than to manufacturing or agriculture. I think it is clear that we have not yet adapted to this change in employment patterns. This lack of adaptation is evident in the kind of vocational training that has been supported in this country, which until this year was largely vocational training of agricultural and home economics workers.

The fourth significant factor has to do with the rate of productivity increase. This is a subject of considerable controversy. Wiser people than I disagree on such questions as: How fast is productivity increasing and what does this mean in terms of the economy? Where has the increased output per worker occurred? How has it depended upon time? What is the character of the present changes in the output per worker? Nonetheless, there is one fact about productivity on which there can be little argument. If we look at the whole of this century, the rate of productivity increase per worker as measured by gross output has not changed remarkably. There have been ups and downs, wars and depressions, but the rate of increase in output per worker as measured by the Gross National Product (which is not a perfect measure of the growth of output of the economy) has not changed greatly during that period.

What this means is that there has not been any underlying change in the rate of the application of new mechanisms, or of automation techniques, to the economy. Moreover, what progress has occurred has been very uneven. It has come in spurts with the peaks coinciding with such special circumstances as wars, or satisfaction of postwar demands, and with periods of stagnation in between. I think, therefore, that any generalizations about increases in output per worker over the last few years, expressed as averages, have very little meaning. The most significant thing is that during the last five years (and until this year) we have not reached the high rates of productivity increases that we reached in other periods of our growth. Since the last war, the rate of increase in output per worker in the agricultural sector has been somewhat higher than in manufacturing, and the latter somewhat higher than in services. Moreover, our rate of increase in productivity in manufacturing has not been as great as in some other countries.

In perspective, then, the problem of the rate of worker displacement does not constitute any more serious a problem today than in the past. This does not mean that we are currently adapting ourselves to that change in productivity as rapidly as we might have in the past, but it does mean that the problem is not any more severe on a percentage basis. On an absolute basis it is, of course, more severe.

It is obvious that improvement in productivity is an important source of improvement in a nation's wealth. A second source of improvement is in the products themselves, their quality. The improvement in quality of a product or service is not measured by our current statistics, but rather by the change in the character of the lives we lead. Unfortunately, we do not know how to measure that change.

My own view is that since 1929 we have been unable, on a steady-state basis, to deal with the problem of adjusting to technological change. The large number of people out of work during the 1930s were ultimately employed through the huge requirements of World War II. Following the war, the pent-up consumer demand kept people at work for a while. The Korean conflict then took up the slack, but not sufficiently to take care of the displacements that were occurring as a result of improvements in productivity and our inadequate training programs. To phrase it differently, we have never solved the key economic problem of our society. We have succeeded temporarily but only through extraordinary governmental intervention, sometimes premeditated, sometimes not.

The most important difficulty with our economy is the continued existence of a significant degree of unemployment. The tax cut will help, but will not fully solve, the problem of unemployment, especially in those areas of our country, such as Appalachia, or in many of our cities, where unemployment is the result of our inability to adapt to technological change and the result of deep social maladjustment.

The problem of unemployment will be alleviated, but again, not fully solved, by a higher level of economic activity. There will still be unskilled workers unable to find jobs; at the same time there are jobs unable to find workers with the required skills. This is because of our long-standing inability to solve the problems of social readjustment and of technological change in a rapidly changing society. This is due in part to inadequate mobility of workers, to inadequate programs of retraining, but, perhaps more than we realize, to the fact that new jobs are not being created by new businesses which produce products that the wealthy consumers of America are willing to buy. The American consumers' demands are becoming more social than private, and tend to concern the social problems of an urban society.

The next characteristic of our society to be noted is the huge expenditures for military and space development. These expenditures are aimed at increasing the "technologically possible," for in the military and space business it is not the relation between cost and benefit to society which is important, but rather the relation between cost and effectiveness, and these are very different things. The military and space effort will always present new opportunities deriving from science and technology, but they will do this without dealing with the complex problem of introducing those new possibilities into a society often unwilling and unable to accept them.

The limitation to the exploration of space or to the development of weapons is the limitation of our technical and scientific capability. There are few social, political, or economic problems directly involved. In the case of meeting civilian needs, on the other hand, the limitations to the application of technology involve such things as political factors, social resistance to change, the cost of the social and economic displacement brought about by the change, the understanding of the people, and the character of the change that has to take place.

Military and space activity, however, determines in large measure the character of advanced technology in America. It strongly influences the character of our advanced educational systems. A substantial part of graduate education in science and engineering, and to an increasing degree in the social sciences, is supported by the military and space effort. The fact that these efforts are sometimes enlightened and do not impose tight restrictions upon universities is not particularly pertinent. The important point is that the expenditures are based on national security needs rather than on scientific exploration, or other social or economic needs.

With the great rise in government research and development expenditures for military, space, and similar special-purpose programs, however, the amount of effort by the private sector of the economy in research and development has not grown as rapidly. In fact the number of people engaged in advanced technical work, supported by industry, for civilian rather than military and space purposes has declined relatively in the last several years. The number of people at universities engaged in those intellectual problems that are connected with our civilian society has also declined during the same period, relatively, if not absolutely.

Private industry has constantly made the claim that its investment in research and development for the space and military effort was justified because it would put industry in a position to exploit this technology for civilian markets in the future. I suspect that this investment yields a good return on its own merits and is therefore made for that reason, rather than because the resulting technology can eventually be translated into civilian productive capabilities. That translation is extraordinarily difficult, for technology that is relevant to the space effort is not particularly pertinent to the nonmilitary, nonspace economy, except possibly in the long term.

I have no doubt at all that the science having to do with quantum mechanics, nuclear physics, and solid-state physics, which today would not be as generously supported as it is if it were not for our military and space effort, will have a long-term effect on our civilian industrial capabilities. I have no doubt, for example, that the support for complicated computers largely to serve military and space requirements shortens the time required for these computers to be developed for civilian purposes. The point I am trying to make is that this is an indirect and expensive way of developing a civilian technology. As someone has said facetiously, "To go to the moon is a hell of a way to get better Quaker Oats," and conversely, to develop a new Quaker Oats is a hell of a way to get to the moon. Nobody would ever pretend that the spillover from civilian industrial research and technology would get us to the moon.

I think it is just as foolhardy to depend upon military and space activity to provide for the civilian needs of our society. This is not to argue that the military and space programs do not support good research and our university system; nor does it deny that there will be useful consequences of our vast military and space research and development programs. But I do argue that this is not a very efficient way to support education *per se*, or to support the development of civilian industry in this country. These important objectives should be supported explicitly and on their own merits just as we should support the military and space programs for their purposes.

The growth and improvement of the private sector of our economy are increasingly limited by nontechnical, social factors. I would like to give you a few examples. The housing and construction industry is by far the single largest industry in America. Roughly speaking, it is a $100 billion industry. It is limited, in technical progress, not because we do not have a technology that can be applied to it, but because we do not have the social structure and understanding to deal with complex social problems, such as the multiplicity of building codes which prevents the introduction of new technology. Systems analysis, computer techniques, operations research, new ways of construction, all could make adequate and cheaper housing more broadly available. We are limited not by our inability to apply science or by a dearth of technical possibilities, but by social, political, and economic considerations.

We spend about $300 million per year to subsidize the American merchant marine; I see no program or plan to reduce that subsidy through the application of modern technology. We are capable of manufacturing ships today that could be operated with an average crew of twenty people. It is not difficult. The Japanese today are operating ships with crews of twenty-five while our average freighter has forty-five crew members. So, in the maritime industry we must devise a way to introduce new technology while taking into account the problems it would create for those who make their living at sea.

Our railroad industry was well aware twenty years ago of the consequences of dieselization. Yet neither labor nor management provided in advance for the necessary retraining and replacement of labor. Nor have the railroads taken full advantage of the potential technology even today. Had we been alert twenty or thirty years ago, I believe we could have met some of the challenges of the technological "end run" executed by the airplane and the truck. In a very real sense the air and truck transport business has succeeded in invading the railroad business because the railroad industry did not (or could not) take advantage of the opportunities which stood before it. Presently the total research and development budget of American railroads is about $7 million a year. This is approximately what it costs to launch a single Tiros meteorological satellite! It must be said, in all fairness, that the problems of the railroad industry are not limited to research and development, but encompass important social and political considerations as well. Again the problem is to deal with the technical and social factors together.

A vital problem, about which there is a great deal of argument, is that of reducing our military program, not so much because it is costly, but because our needs are changing. I have already pointed out that military expenditures put many people to work. What to do with them if the program were to be reduced is a serious matter.

In some ways, the problem of adjusting the economy to cope with substantially reduced military spending is as complex as providing for security itself. There is the danger that areas of our country now dependent upon government contracts may take the short-range view of resisting reduced programs, and that other areas, not so dependent, may take the similarly short-sighted view of not wishing to see the blow eased. The resistance to a reduction of military expenditures often comes from Congressmen's understandable fears of the economic consequences back home.

Take California as an example—40 percent of its work force are directly or indirectly dependent upon military production and space research and development. Next year, for the first time since 1946, this figure is not going to increase, and the number of workers supported by the federal government in California will remain constant. There is some unemployment on the San Francisco peninsula already. This is unemployment of a highly educated, highly trained, highly vocal, and highly, though narrowly, skilled group of technical people. Here we have technological unemployment of a very serious kind, a fact we have to face in viewing technology and the national interest. A similar potential problem faces the Route 12-8 region outside Boston and other areas of the country.

What should an attitude toward this problem be? What

public policy should we adopt? The nation has survived similar events in the past. And, after all, we are only speaking of about $2 or $3 billion of research and development out of a $600 billion plus economy. Furthermore, it can be argued that the government has only a limited obligation to assure the survival of every corporate enterprise in the country. Against these points, however, we must realize that the $2 or $3 billion of research and development is a large part of the total performed in the country. In this light, a program of retraining for these highly skilled people would surely be in the public interest.

The next factor we have to consider is that although our problems are getting more complex, the power of science is also growing. Increasingly we are producing machinery which can deal with complex systems and handle the vast amount of data that must be considered in making plans for a city rather than a house, for a transportation network rather than a road. Perhaps this advance in man's capabilities, coming heavily from his investment in military and space programs, may be the most significant spillover—if we can but learn to use it to deal with the complexities of modern society.

Another factor that must be considered is how to use advanced technology in the less-developed countries. The problem here is not one of providing these nations with highly technical education before they know what to do with it. Nor is the problem necessarily one of providing on a country-to-country, government-to-government basis the social overhead or infrastructure required for the future development of their economies. Rather, it is the problem of developing an institution-to-institution relationship, creating a large number of separate points in a less-developed country where jobs can be created and local industries developed, where the techniques of our society can be adapted to local industrial needs. As long as we tend to make our aid and contributions on a government-to-government basis, with the wasteful techniques which that can encourage, the problem of adapting our technology to meet the needs of the less-developed nations is going to remain a troublesome one. The need is for the ingenious adaptation of techniques to local surroundings, and to do this requires a high quality of mind and a special kind of inventiveness.

These, I think, are some of the factors that must be kept in mind in trying to understand how science can be applied to the civilian sector of the economy as distinguished from the military and space efforts. The problem in the civilian sector is to get ingenious adaptation and invention. This is not necessarily done by the fellow who is trained to do conventional research and development, not because his ingenuity is any less but because he has learned a different technique.

What, then, are the characteristics of technological change? How does it come about? I think the single most important characteristic is that it generally comes about by invasion. Minor improvements, like modifications of a dishwasher, do not come about in this way, but major technological changes do.

America's new leadership in the photographic industry, the Polaroid Company for example, derives from the fact that an ingenious inventor with little formal scientific education invaded the domain of the large, established domestic companies, as well as the earlier, cheap but obviously effective Japanese optical systems (which are now very good). Similarly, Richard Morse in 1942 invaded the vacuum system business, establishing the National Research Corporation. Morse used vacuum techniques for the manufacture of concentrated orange juice, and initiated this industry. Another example is the invasion of the machine tool business by other techniques, such as tungsten carbides from Germany which revolutionized the American machine-tool business.

Since major technological change often results from an invasion by outside parties, we must encourage the invaders if we want to stimulate innovation. To do so requires that technological information must be disseminated throughout the land to the maximum number of people who can use it. Venture capital must be encouraged and risks must be shared by society, because the introduction of change is a risky business and all the benefit does not necessarily fall to the man who initiates it.

I think most people agree that science must be supported by society, because its benefits are diffuse. It is less clearly understood that practically all the major new technologies which have been developed in the past were also socially supported. Aeronautical technology, the building of the railroads in the West— these had to be social investments because no single person or enterprise could possibly capture all the benefits. The private inventor, however, is still an immensely important individual, much more important than most people realize, even when compared to the huge industrial research organizations created to develop new products.

Another factor which must be considered is the disenchantment of many big businesses with science as a major, direct avenue to new products. This does not mean that science is not pervasive or not extraordinarily important for our society, but rather that the direct application of science is a different enterprise from the research which leads to new products. Nor is this to say that scientists cannot develop new products, particularly when the science is new. Atomic power is an example of just that sort of thing. What I want to emphasize is that the benefit of the new science does not often accrue to the company or the individual who makes the investment.

This leads me to the conclusion that the two most important limitations on the advance of technology in the civilian sector are the absence of people who understand both technology and science on the one hand, and the absence of people who understand the business aspect of a technological enterprise, that is, the risk-taking and the financing, on the other. I would like to make it clear that the entrepreneur not only has to invent the product and find a market for it, he also has to invent a method to sell it, and sometimes even invent the customer.

The latter is frequently very important. The electric power industry in America was created because its developers lent the power-producing companies the money to buy the equipment the developers produced. Not only that, they developed appliances which used the power which used the equipment which enabled them to make money. Thus there were not only scientific and technical possibilities, but also wise men who knew something about science and its possibilities on the one hand, and about the character of the market place on the other.

Unfortunately, the number of individuals who combine these talents in their own person is quite limited. This person doesn't have to be a solitary inventor working in a garret. It may be someone working with other people in a complex technical organization. But both types—the inventor who conceives a product or service that will meet a customer's needs, whether that customer is an individual or a society, and the entrepreneur who underwrites and markets the new thing—are absolutely essential for developing the civilian technology to meet our society's complex needs.

A third limitation is one to which I have already alluded. It is that we do not know how to deal with the displacements that technology is bound to cause: the individual displacement as in the case of elevator operators, the regional displacement as in the case of Appalachia, or the industrial displacement as in the case of the textile industry. All too often the individual worker, company, or region has to bear the cost of technological change, while the rest of society reaps the benefits.

The three major limitations, therefore, on technological change are: 1) the lack of inventive talent; 2) the lack of entrepreneurial talent; 3) the high social cost of technological change.

What are the civilian needs that must be met by science and

technology? It seems to me that right now the most important need is to create new jobs. Frankly, I don't know how to do that in the middle of New York City or any other specific locale. I have thought about it a great deal, and while I have come to no fixed conclusions these are some of the points to keep in mind. It is important, first of all, to provide generous assistance to the entrepreneur in the local area who is creating the jobs. We should consider special tax credits and faster write-offs, rather than subsidies to encourage new business.

I believe, however, that education comes after job creation, and not before. These things are symbiotic, it is true. Obviously we cannot afford to educate people unless we have people working to produce goods, and we cannot produce the higher valued goods unless the educational system is highly developed. Still, the first step is to create the jobs.

In the case of an urban, poverty-stricken community, I think that we might somehow shelter new industry for a while. Obviously a new industry cannot afford to pay the same wages as a more firmly established industry—at least, not at the beginning. We might need exemption from some federal laws, minimum wage, possibly others. Maybe we should put the people to work repairing their own buildings, making their own materials. Certainly the problems of regional unemployment are very complicated, involving as they do complex political, social, and economic factors. Nevertheless, the single most important need to which technology can be applied is the creation of new jobs.

The second important national need is city-building. This offers the single most important business opportunity in America today. A product that many people would like to buy is a place to live that has recreational, cultural, and industrial opportunities, a place where parents can raise their children and have them properly educated, that is, cities ranging in size from 50,000 to 100,000 people. Yet we don't have anyone trying to supply this potential market.

The third need is in the field of education. Obviously our education must be continuous throughout life and an integral part of the local community. It must also apply the techniques that we already have for handling and disseminating information.

Another need is to apply our advancing technology itself to decision-making, at the city, state, and most important, the national level. We simply do not use technology adequately for decision-making. We do not, for example, simulate the economy in a national way. Economists now make shrewd guesses about the effects of tax reduction, but they shouldn't have to guess. We ought to be able to apply systems analysis to complicated economic and social problems. If we don't, we are not going to be able to deal with the more complicated questions of society in the future.

The last need, which to me happens to be the most important, is that of securing a way to disarmament, a way to use technology for arms inspection, and for the conversion of the people previously engaged in defense activity to the task of solving the complex problems of an urban society—city-building and urban development.

The character of the problems that face technology and its application to civilian activities clearly requires a deep understanding of science and its application (and a support for science on its own merits—as an item of faith). It also requires a group of people with a deep understanding of economics, politics, and social science. Otherwise we will be unable to apply science and advanced technology to the problems of a modern, complex, urban society.

DISCUSSION

UNEMPLOYMENT AND JOB CREATION

Question: Can more jobs be created through expansion of the service industries?

Mr. Hollomon: In the service industries the problem of job creation is not primarily technological. It is rather one of discovering ways to enable a service worker to offer his services at what the consumer considers an acceptable price. There are large regions of the country that are in different states of economic development from others. If these regions were separate countries they would be able to enact their own rules and regulations. Differing wage rates, tariffs, and subsidies would all furnish ways to create jobs. We, however, are a single nation. From the point of view of creating a single, large market this is desirable, but it makes it difficult for us to take into account differences in local conditions.

Comment: In New York we have people out of work when there are jobs to be done. For example, we suffer from inadequate mail service; yet there are unemployed people who theoretically could be used to increase the frequency of deliveries.

Mr. Hollomon: The cleaning of houses and offices is another area where employment could be expanded, but the price of commercial business cleaning in New York City is $1.75 an hour. Yet, at the same time, the social cost of unemployment is enormous.

Question: Do you think there would be jobs for the scientific and technical personnel in Palo Alto if their salaries were cut?

Mr. Hollomon: There is no other alternative unless we want to perpetually subsidize the people who went into military and space development work. Like any other worker, the unemployed scientist or technician must be retrained. Part of this group will go back into sales and manufacturing activity. Compared to the prewar period, the fraction of our scientists and engineers that performs research and development today rather than sales and manufacturing is enormous. Their return to these other types of activities would be beneficial, but at present salaries it would be difficult for them to find such jobs. Moreover, they are going to have to adjust their skills to new situations. They haven't been worried about costs and things like that in space and military technology. They have been concerned only about what is technically possible.

Question: How does another highly industrial society, the Soviet Union, handle the problem of unemployment and job security?

Comment: There is first of all the question of whether the Soviet Union does have unemployment. The answer is yes. Part of the unemployment is hidden out on the farms, but there is also unemployment in the cities.

Mr. Hollomon: Also the Soviet Union is a society which has a peculiar advantage, in that they have something to copy, the United States. They know what kind of skills are going to be needed, at least for the near future, and they can train for these skills. This is not unlike the situation which prevailed in America for many years when we could copy the more advanced European technology. For example, most of my graduate teachers received their advanced education in Europe. I suspect that today the reverse is true. Instead of one out of ten, it is nine out of ten who are educated in America. When a country is in the position of being a follower technologically, the problem is entirely different from what it is when a country is in the position of being a leader. This is why the problems of the Soviet Union are different from ours. I do know that they train people very narrowly because they know, roughly speaking, that they will need at least that number of specialized people. They know, too, that they will need these specialized skills for a long time because they are behind other countries.

I don't think that we are going to be able to depend solely upon the private sector to create all the jobs we will need. We have not been able to depend entirely on the private sector for several decades now. However, I think that we can do more in the private sector than we have done. At the same time I think that much more has to be done jointly and cooperatively between the government and the private sector to create new jobs.

Comment: We should be afraid of unemployment, of idle people, not of over-employment, for in the latter case at least the worker has some chance of retaining and developing skills. We should really examine the sort of thing we are doing in the post office, for example, trying to cut down the amount of labor needed through automating jobs, when there are all sorts of tasks that can be performed by workers who possess few skills. We are keeping people idle in part because we have an unbalanced idea of efficiency. We are not looking at the total social system and the total social values. Our unemployment problem is small, from a percentage point of view. But it is possible we may have a high rate of unemployment among highly skilled people, people who are highly vocal, people who could be revolutionaries and upset the whole society if we don't find jobs for them.

Question: What, then, do you do until you figure out a long-range solution? Could you possibly suggest some specific things to create jobs in New York?

Mr. Hollomon: I can say something about the principles involved. For New York you obviously have to create labor-intensive work. Now there are two kinds of labor-intensive work. One is in the construction industry and the other is in the various service industries. My guess is that a large public works program may have to be part of the solution.

Comment: One way of reducing unemployment in New York City lies partly in the mass use of relatively unskilled labor in construction work. It requires that we free ourselves of an attitude that is partly a fiction and partly a fact, that construction requires skilled labor. If you go to underdeveloped countries you find that they build very good schools, hospitals, and roads for the most part with unskilled labor. If we were to rethink the way construction is carried out in a big city, making use of unskilled labor, making use of our manufacturing skills, instead of the skills of a handicraftsman on the job, and if we were to think in terms of the wage rates of unskilled labor instead of the present wage rates of skilled workers, we might accomplish a tremendous amount.

Comment: There is really a political precondition for being able to do something about this problem. Unless the country decides that the question of unemployment has a much higher priority than it now has, nothing will work in New York or anywhere else.

Mr. Hollomon: It is true that you need the political commitment to do something about unemployment, but you will still have to shelter some of the employment created. You might have to say, "Regular labor with regular wage rates operates generally in this area and another type of labor with lower wage rates is allowed to operate in this other area" because of special considerations such as I described.

UNEMPLOYMENT 1929 AND AFTER

Question: You said that our difficulties seem to date from 1929, but that at the same time the rate of increase in productivity has remained fundamentally the same since 1870. Why do you think that the problem of job creation has intensified since 1929? Has automation or advancing technology been a factor?

Mr. Hollomon: I think that is just catching up with us. We have not achieved what could be called a steady-state solution to the problem. We have improvised, we have benefited from unforeseen circumstances, and these have made it appear at times that the problem has been solved. First there were the large wartime expenditures. Then, after the war there was the pent up consumer demand and the rebuilding of Europe which helped to maintain high levels of employment. But now the rebuilding of Europe has been accomplished and military expenditures are leveling off. The relative amount of social investment by government has decreased. As a result the problem of unemployment is emerging more clearly. I don't think automation has contributed more in recent times than in the past to make the problem any more severe.

Comment: The problem of full employment from an economist's point of view is the cost which such a policy would entail. What you are likely to get are real balance of payments problems, and a substantial increase in price levels. As soon as the economy moves toward full employment, the labor movement becomes more aggressive, which inevitably means a more rapid rise in prices. It is essential, as I see it, to look at more than the particular problems of the central city or Appalachia. Even Appalachia had no substantial problem until unemployment became generally widespread. The first impact of the reduced demand for bituminous coal was in fact absorbed because people could still obtain jobs outside the region. A major portion of the problem in pursuing a policy of full employment is how to handle the side effects. A small amount of unemployment, from a practical point of view, may be better than a rise in the price level, since a rise in the price level affects everybody, while unemployment affects only a small number, and those persons may not have much power at the polls.

Comment: Economic growth, employment, and profits are alternative criteria for the success of an economy. These criteria used to run closely together, but I think that increasingly this is no longer so true. Over the past two years the Gross National Product has grown considerably, 20 percent on a $500 billion base. But employment increased by only two and a half million jobs, leaving millions still unemployed. So we must realize that these are now much more complicated interrelations than they used to be.

Historically we have concealed some of our problems on the farm. A person may be half hungry on the farm but at least he has a house. He may only work a few days during the year, but who is concerned? The Negro problem is partially derived from the fact that the country has been transformed from a predominantly rural to an overwhelmingly urbanized society. The old problems are still with us but in an urban setting they take on a new order of magnitude.

Comment: You have emphasized the importance of increasing mobility but as far as I understand it America has the greatest mobility of any nation in the world.

Mr. Hollomon: But the mobility of people with a single skill after the age of thirty-five, such as is true of many people in Appalachia, is very slight. One problem is that so many Americans now own their own houses that when a single-industry community loses its industry no one wants to move.

THE ENTREPRENEUR: INVASION, INNOVATION, INVENTION, AND SOCIAL COST

Comment: I think the most important problem is the lack of sufficient number of entrepreneurs. We must get some of the new ideas, and I am convinced we have an abundance of them in this country, into the public domain. The problem in this country is not so much to create new inventions. The problem is to get the inventions implemented. Some system of incentives, or some substitute for incentives, to encourage entrepreneurs to put new products on the market is what we need. Most major companies today have inventions that never reach the public. They are not utilized because the companies are afraid to introduce these products.

Mr. Hollomon: Certainly there is ambivalence in the large company. On the one hand, the top management says, "We have to have innovation to survive"; on the other hand it says, "Oh, but look what it does to us. We have to change our manufacturing techniques, our salesmen, our management, everything." But I would say that it would be extraordinary if an invention which comes about through a marriage of what is possible with what is needed happened to come about just at the exact moment when the society is willing to buy it. The invention tends to come before society needs it or is willing to buy it. Therefore you usually have to wait for the invention to become saleable to the public. More frequently than not you have to wait a very long time. Few inventions are made that a society is willing to pay for immediately. The transistor is an example of this. An interesting article in *Science* claims that it was first invented in 1932. In any case the transistor was invented privately, but its time of application was enormously shortened by government subsidies. It is not at all certain that it was done in the most economical fashion. After all, the transistor is a substitute for the vacuum tube and it is not at all certain that society has benefited by pushing the development of the transistor along at the speed with which it has in fact been developed.

What I am saying is that you must wait until society is ready to pay for the inventions. The jet engine is an example. The magnetron tube is an example. Of course, there are some inventions whose use should not be forced. It would be too costly to society if they were. For example, I believe that it would cost this country enormous sums to develop the supersonic transport by 1970, though this would be possible. If we are willing to wait a little longer, I am sure that the market will be big enough and the technology sufficiently advanced so that society would benefit to a greater degree from the development of the supersonic transport.

Comment: Who is the invader? He is probably an inventor within one of these large companies who believes that what he has discovered is a significant, new product but who cannot sell the idea to his top management. The invention sits on the shelf because the company doesn't see fit to innovate. If there were some means whereby this inventor could take his discovery to another company, the innovation might take place.

Mr. Hollomon: I agree with you that it is a problem, but I don't think that it is a crucial problem. I believe that the crucial problem is to provide an atmosphere in which what is called "patient money" will be forthcoming to support the innovation.

Comment: If we say there is something wrong with entrepreneurship, we are perhaps saying that there is something wrong with the basic structure of the corporation in America.

Comment: In this same vein, invasions were probably more successful when people were ignorant of their implications. When people are aware of what may happen, they may become fearful. There may be more of these fearful persons than there are those who see the immediate personal benefit of the invasion. The argument in favor of the invasion is based upon some conception of the total social benefit to be derived by society in the long run, but the immediate problem is that we don't know how to deal with the costs of dislocation, except through such legal and political expedients as the maritime subsidies.

Mr. Hollomon: That is why the problem of assuring that the adjustment takes place without any one individual having to bear the full cost is so crucial.

Comment: But it is not just an economic question. It may be difficult to get people to welcome change in their personal habits and patterns of life.

Mr. Hollomon: The most important thing is to establish the fact that technological change is going to happen and that it is desirable in the long term. Many people already understand that. Surprisingly, people in the labor movement sometimes understand best that technological change is both essential and desirable for the well-being of society, particularly for the poorer part of society. But there must also be techniques to deal with the problems of adjustment.

Question: Was the complex of industries that centered on Route 128 outside Boston a development sufficiently controlled and managed by one group so that it could benefit from its mistakes when it decided to create another Route 128? It sounds as if this is what you had in mind.

Mr. Hollomon: I was not likening what happened on Boston's Route 128 to city building. But I do believe that the people directly interested in Boston's economic development have learned something from this lesson, and they are now trying to diversify the economic base. I think that the same thing has happened in Palo Alto. The fact that those two areas have been cited as examples of how to develop industries based upon science leads me to observe that we cannot generalize on the basis of these two regions because their industries are largely based on military and space efforts.

SOCIAL RESEARCH AND SOCIAL INVENTIONS

Comment: The burden of your remarks is that we need social inventions and social technology to furnish us with the ability to change institutions and people, and that to create these new social institutions we first need a better understanding of people and institutions. It seems to me that you could say that we already know enough about people and institutions but that we haven't learned to use this knowledge. If not, why not? Or you could say, for example, that we really don't know enough to build the kind of cities you suggest we would like to have. In that case we need a large investment in the development of social technology. Or you can assume that we will simply bungle through, depending upon the conventional social, economic, and political wisdom.

Mr. Hollomon: We have not been so unsuccessful in applying social knowledge as you seem to imply. In my view we have been enormously successful in agriculture. We have also been successful in parts of our educational system, in science and engineering. Now, to build successful cities we need to have trials. We need to learn by doing. I am not sure we will ever understand enough about how to build a city. We have to go ahead and build them, then adjust them, because that is the way any complicated engineering or technological project or social institution is created. I don't think we have yet learned to use the strength of our educational institutions to help solve the problems of local communities. More than at any time in the past we must depend upon technical, scientific, and intellectual rather than natural resources to provide the basis for our social development. Research in the social sciences will not hurt us, but I am not convinced that it will help us in solving immediate problems. More important, I think, is to get people into the political arena who want to work on these problems.

Comment: The speaker has raised the problem of creating cities. But creating a city is not as mechanical a job as he implies. It is a biological problem and one cannot force biology upon society.

Question: Do you think that the political tools available to us, and the institutional procedures we presently use, are enough to overcome the kinds of inertia that you have described?

Mr. Hollomon: There are many problems where the obstacles are not lack of knowledge but rather much more the problem of getting the commitment of people, the commitment of the country to do something about the problem. People can say, "I want to do social research," or they can say, "I want to get into the political arena and do something about the problem." I think that the latter is the most important thing today. One could draw an analogy in science and technology with some people wanting to discover some unknown fact of nature and someone else wanting to bring a new idea into the market place.

Question: How do we get social inventions into the political arena so that we can try them?

Mr. Hollomon: It is possible with our existing political machinery to get interested, intelligent, dedicated people to operate in the political arena to help make social experiments in a clinical fashion, but it is only possible if there is a prior political commitment from the people to want to do it.

Question: You have said that we need to use analysis and operations research for these problems. But these techniques need a reasonably rigorous set of inputs if they are to be used to provide models for dealing with social problems. Can you in fact simulate the economy today?

Mr. Hollomon: We can simulate it well enough so that we can make relatively intelligent decisions about tax reduction. If we put the kind of effort into simulation to answer social questions that we put into other kinds of technical problems, we will be able to make predictions substantially better than is now being done. We can get detailed answers to specific questions about the economy, the character of transportation, the movement of goods, and so forth.

Question: I would like to raise a more general question. What disturbs me as we have talked today and upon a number of other occasions is that we keep on discussing how we can keep the system going for another two or three or four or five years. Now if it is possible that in ten or twenty years we will have a cybernetic era, shouldn't we as a seminar face this issue?

Mr. Hollomon: I see no evidence for this and I know no one who has deeply thought about the question who sees any major change of this kind taking place.

Technology and Social Change: A Congressman's View

by JOHN BRADEMAS

Congressman from Indiana

I AM GOING TO SPEAK from the personal perspective, if I may, of a congressman who comes from a particular part of the United States, from a particular congressional district, who belongs to a particular political party and to a particular segment of that party, who has an interest in education and serves on a committee that gives him an opportunity to give some attention to these interests—the Committee on Education and Labor. I shall discuss my roles as a congressman in relation to the problems occasioned by the development of science and technology in the United States.

One of the problems many laymen—especially scientists and people in the university community—face when they look at Congress is an inadequate appreciation of the work load carried by a member of the United States House of Representatives or Senate. They are not aware that a congressman is many people. I represent a farm-factory district of about half a million people. My job, like that of my colleagues, is really several jobs. I'm a lawyer for the people of my district in their dealings with the various executive agencies. I'm an ambassador from my district trying to bring good things back to the people I represent in the form, for example, of defense business or, right now, manpower retraining programs. Other congressmen emphasize what are commonly called pork barrel projects, especially public works projects. The new pork barrel about which I want to say something a little later is represented by science installations. In my judgment, a congressman must also, if he is going to do his job in a responsible way, try to be an educator and not simply to mirror the views of the people in his district. And of course I have to be a politician since I have to run for reelection every two years from a highly marginal district.

The complexity of this job means that congressmen do not have nearly the time they need to devote to science and similar matters. It isn't that congressmen are deliberately ignorant about these things; they just don't have enough time to devote to them.

There is a second reason we are not as well equipped to deal with the impact of science and technology in our country as we'd like to be—our own lack of expertise. I think I'm right in saying that there is not a single scientist, in the sense in which we normally use that word, in either the House or the Senate. Now it may be that scientists don't know how to get elected, but I rather imagine that it's because scientists don't really want to get elected; that the scientist and politician are two different kinds of animals, and that where we politicians have a vested interest in ambiguity, the scientist has a vested interest in precision. The scientist is apt to become irritated when we politicians deliberately try to cloud things up. But that's the nature of the political animal, and it often serves (upon reflection) a useful purpose.

Some of the recent developments, such as the establishment of the Daddario and the Elliott select committees on research,[1] indicate that there is an increasing uneasiness among members of

[1] Emilio Daddario, Democrat of Connecticut, chairman of the Subcommittee on Science Research and Development, and Carl Elliott, Democrat of Alabama, former chairman of the House Select Committee to Investigate Government Research Activities.

Congress because they feel they must rely far too much on the executive branch of government for expertise, that they have to take much too much on faith when they vote those billions of dollars for scientific research.

Then there's another problem—the fact that authority over science programs at the federal level is spread among a wide variety of congressional committees. I serve on the Education Committee, but we don't have anything to say about the National Science Foundation programs. That's handled by the Science and Astronautics Committee. The Science Committee, on the other hand, doesn't have anything to say about college aid bills (such as the one passed in 1963). Now, there's no simple resolution of these conflicts, but it does mean that it's difficult for us to get any kind of overall or coordinated view of the impact of science, at least so far as federal legislation is concerned.

Because there were two or three pieces of legislation dealing with science and education matters that I was interested in, and because I felt that I could get little help from my committee staff, I once put an engineering dean from Purdue on my own staff full time for several months so that I could have somebody in the office to talk to about these subjects. But this is not the orthodox procedure. It is a technique that I think might fruitfully be explored, for I think it might well serve the interests of large universities who care how federal monies are spent for education and science to let one of their top people go to Washington to work in the office of a congressman for a while.

As you may know, legislation has been introduced that would provide for the establishment of a congressional office of science and technology. My own view is that this proposal is not likely to get very far. It's a kind of counterpart to Jerome Wiesner's operation, to describe it simply.[2] It won't get far because

[2] Former Special Assistant to the President for Scientific Matters.

we in Congress like to take our help and advice from men loyal to us. In an office serving the whole Congress the loyalties would be too diffuse to satisfy the individual congressman. Wiesner knows for whom he works; there's no question about that. It would be quite different if a science adviser operation were set up on Capitol Hill.

Information may be available to us soon from another source. The American Association for the Advancement of Science, which now publishes *Science*, is considering publishing shortly a journal devoted primarily to the relationships between government and science.

Let me turn to some specific substantive problems that concern a congressman who's interested in the impact of science and technology. Obviously as important as any other is their effect on his own local economy. With the defense budget at $50 billion, today's congressmen may often have to become a kind of lobbyist for bringing more defense business to his own district, especially if he has a district like mine, and in most districts where there's any kind of an urban population the government is spending defense money. This raises certain important questions: Is this an appropriate role for a congressman? And how effective is this kind of effort?

The Studebaker situation is perhaps a good example. The company's plant was located in South Bend, Indiana, in my district. I can say that the Indiana senators and I, in the years that I've been a member of Congress, worked very hard on behalf of the Studebaker corporation. Since 1961 it received about $200 million in government contracts, and we continually fought to bring Studebaker to the attention of procurement officials in the Pentagon.

We're often asked by newsmen, "Are you suggesting that political influence has something to do with the allocation of these funds?" It's much more complicated than that. The fact that I am a Democrat during a Democratic administration doesn't mean that I always get the defense contracts for my district, because other congressmen are fighting for companies in their districts. I never know with complete assurance, when the contract is finally awarded, whether my effort made the difference or not. In any event, to be very candid about it, the effort has to be made because the public expects it.

Let me turn to another problem that's directly related to this one, the relationship between defense production procurement and Defense Department research-and-development expenditures. A year or so ago the then Under-Secretary of Defense, Roswell Gilpatric authorized publication of a report entitled, "The Changing Patterns of Defense Procurement," which created quite a stir in Washington. The point of this report was that wherever federal research-and-development dollars go, defense production contracts are very likely to follow. And where do the federal research dollars go? To those parts of the United States which have the great university research centers and the highly trained scientific and technical personnel, which means California, Massachusetts, and now, increasingly, Texas. I remember sitting one day with Senators Hubert Humphrey and Eugene McCarthy of Minnesota, the two senators from Indiana, Vance Hartke and Birch Bayh, and some top Pentagon officials discussing whether Studebaker was going to get a big contract or whether it would go to some California firm. Senator Humphrey warned the Defense Department officials, "You know, if you keep putting this money into the East and West coasts, you're going to make this country lopsided." This is precisely what is now happening. If you were to take a map and shade it in terms of where federal research monies go, the point would be obvious.

I, myself, don't take the attitude that a lot of Midwest congressmen do—perhaps for public consumption—that we ought to assume that we're being cheated in some way by the people in the Pentagon. My feeling is that we have to lobby for our part of the United States, but more important, we have to begin building up our own university research centers, doing what we can to get back into the economic ballgame. For example, recently I accompanied a group of community leaders from South Bend to testify before officials of the National Aeronautics and Space Administration on behalf of the University of Notre Dame as the site for the proposed $56 million NASA electronics research center, which, as you know, some people say is already destined to go to Boston.[3] Later I saw Congressman Charles Halleck, the House Republican leader, who would like to see the center located at Purdue University, which happens to be in his own congressional district. He said, "Do you really think you can get that in Indiana?"

I said, "I'm not sure, Charlie, but I'll tell you this, I'll vote with you on a roll call against Boston."

I've talked to a lot of my Midwest colleagues and we're getting our backs up on this kind of thing. How can our communities ever break out of the circle if strength follows strength?

In my judgment, therefore, one of a congressman's jobs is to go back home and emphasize to the people of his district—especially the university and industrial community—that they will have to work more closely together than they have in the past if they want to win federal research-and-development contracts. Even Notre Dame's Father Hesburgh, who's an extraordinarily imaginative leader in American education, was reluctant for many years to commit his university to working closely with South Bend business leaders to meet some of the industrial needs of the local community. But then a year or so ago I organized a conference at Notre Dame, bringing in people from Bendix, Studebaker, and other firms, to encourage some long-range pub-

---

[3] Boston was, in fact, later selected as the site of the NASA electronics research center.

lic thinking about this problem. I'm glad to say the attitudes in my area are changing.

Another example of the kind of problem that comes to the attention of Congress in considering the impact of science and technology is one with which the Civilian Technology Program, which Mr. Hollomon has been pushing, would attempt to deal. The federal government now spends vast sums of money on research and development, but this money goes primarily into the areas of atomic energy, space, defense, public health, and agriculture. Nor does private industry spend more than a very small fraction of its research dollar on enhancing the productivity of the civilian sector of the economy. The point is that we spend a lot of money for research and development in this country but we don't spend very much of it to increase the Gross National Product, the constant increase of which makes it possible to spend so much money on research and development for these other, essentially nonproductive items. For these reasons, Wiesner and Hollomon put together, with President Kennedy's strong support, the Civilian Technology Program, designed, among other objectives, to establish university and industry extension centers along the lines of the land grant agricultural extension system which has done so much to enhance the productivity of American farms.

But in attempting to establish such a program, one very quickly runs into stone walls. There are some industries that simply do not want to increase their productivity. They like things as they are. The construction industry is an example. The building trades contractors are in league with the building trades unions. They don't want to know how to make bricks faster or better or cheaper. A close look at the makeup of the Appropriations Committee of the House of Representatives would show that the key figures on the subcommittee that deals with appropriations for the Department of Commerce are very close to powerful lobbyists for the Clay Pipe Institute and similar trade associations. It is easy to realize why it is that when the Department of Commerce appropriations bill reaches the floor of the House of Representatives it contains no money to improve the productivity of the building trades industry. This is another example of the kind of problem one runs into in Congress.

Let me now comment on legislation that deals with science and technology. I think it's fair to say that the 88th Congress made one of the finest records on behalf of American education of any Congress in history. First of all, there was the medical school bill. Parenthetically, let me point out that this bill came out of the Commerce Committee, not the Education Committee. Why? Because for many reasons it's often easier to pass an education bill if it comes out of the Commerce Committee than if it comes out of the Education Committee. It addition to the medical school bill, there was the college aid bill, which provides construction grants for church-supported as well as public colleges and universities, but for categorical, not general purposes. The money goes to provide assistance in building facilities to teach science, engineering, modern foreign languages, and others that don't involve religion in any way.

Also there's the Manpower Act, which was amended in important ways. I'll give you an example of the kind of problem that we have run into in my home community which the amendments seek to remedy. One of the chief reasons persons in South Bend, Indiana, and other parts of the northern United States otherwise qualified for vocational training could not take advantage of the manpower retraining program was that they were illiterate. The Manpower bill attempts to meet this problem by making it possible for those persons to receive limited financial assistance while they learn to read and write.

I must say a word about the vocational education bill. I think that passage of this bill represented the most significant advance in vocational education since the federal program was first begun under President Wilson. We've drastically modernized the program trying, over the cries and objections of many vocational educators—not all of them—to bring it into line with mid-twentieth-century conditions. This bill finally recognizes the fact that everybody in this country doesn't live on farms, that some people live in cities and need a vocational education program geared to urban needs.

Now, what are some of the principal political problems encountered when it comes to federal aid to education? Clearly one of the most difficult is religion, which still remains the chief stumbling block to the passage of a federal aid to elementary and secondary education bill. We finally overcame the problem in the field of higher education but it took several years to do it. Even then we had to make the assistance specific, based on certain categories, rather than general.

Civil rights is another problem. We never know when it is going to be a barrier to the passage of an education bill. If we had tried to pass the vocational education bill in the summer of 1963, and somebody had offered a civil rights rider, that bill would have been in deep touble. But in the fall of 1963 nobody offered that civil rights rider. You may wonder why. The answer is that in the intervening period the temper of the country changed a great deal on this particular issue. As a consequence the barrier was removed. But who knows what may happen to a federal aid to education bill next year?

One of the problems I have often encountered in this area is the narrowness of some of the lobbyists—not only lobbyists in the usual sense of organizations centered in Washington which speak for specific interest groups, but also people in general—who have real interest in education legislation. It's very irritating to see how some vocational educators don't care if Congress passes a college aid bill; and the college aid people don't care if it passes a vocational education bill. One hopes now that the passage of all this legislation will moderate some of the intensely partisan feelings that crippled our efforts to pass such legislation in this field in the past. There is one factor now that isn't as much of a problem as it was when I first came to Congress in 1959 and that is ideology. In other words, there aren't as many people saying, "Federal aid to education is iniquitous." There's just too much federal aid to education already on the books to be able to make a very persuasive argument in that respect.

The final problem, and I think it is a serious one, is apathy. I find, curiously enough, that as a politician I have to spend a lot of my time chasing around the United States making speeches to educators urging them to lobby in support of legislation that will help education. To paraphrase something Norman Cousins once wrote, we have a problem in this country of the default of educated man when it comes to supporting education legislation. For some reason, otherwise intelligent and educated people think that it makes no difference whether they tell their congressman how they feel about these matters, when the fact is that he cares very much, especially since they're usually informed, articulate persons.

Having said all that, let me make just one more point, because I think it bears directly on the relationship between education and science. It seems to me that we in Congress who vote all this money for scientific enterprise are going to have to begin thinking in terms not only of dollar budgets for such programs, but in terms of manpower budgets as well. We vote money for these vast programs on the blithe assumption that the highly trained professional and scientific personnel required to perform this research will drop out of the skies. That, as you know, is not the way the world is. And I suppose that final remark is the justification I offer for having talked so much about education.

DISCUSSION

THE NATURAL SCIENCES

Comment: You spoke of the role of Congress as a whole vis-à-vis the executive branch in obtaining scientific advice and playing a significant role in decision-making. But what strikes an observer is that the people who are taking the lead in asserting a greater voice for Congress in these matters have an animus against science.

Mr. Brademas: I'd have to quarrel with your presupposition. What scientists often do is to mistake congressional interest, and especially this sudden ballooning congressional interest, for hostility to science. I talked to Carl Elliott about his committee and he said, "Now we will be accused"—this was before he formed his committee—"by the scientific community of trying to do science in, but those of us who are really friends of research had better start looking into some of the problems that may have developed in this area now, before troubles develop later on that will ultimately cause the scientific community a lot more harm." This was his justification for the investigation, and an indication of his attitude toward science.

Congressman Daddario's committee has put out a publication called "Government and Science." It's a statement of purpose. I think if you were to read through this, you would be very happy to have a man like Daddario heading up a subcommittee investigating government support of scientific research. I think if you were to look through this document, you would understand he has been moved by genuine concern of the sort I've indicated, and that he has no hostility toward science as such. Indeed, one of the points that Representative Elliott made when he opened his hearings was that research is like motherhood. Everybody is for it; nobody's against it. How can anyone really make a case that congressmen are hostile to research? We're just curious to know how those funds are being spent. Right now, we don't know, or at least don't know enough.

Question: I wonder if you would follow up these remarks that you made about Congress and the problem of scientific expertise. You posit as an almost insoluble dilemma in the sense that there are many committees in Congress concerned with science in various ways; yet a single congressional office, designed to eliminate duplication, wouldn't do the trick because it wouldn't be loyal to any one of you.

Mr. Brademas: First of all, I should note that Brookings Institution has tried to tackle this problem. They set up a series of round tables for members of Congress to which were invited distinguished scientists who spoke on their particular disciplines; and I think the experiment was an extremely useful one. I have to say at the same time, however, that it really doesn't answer the fundamental problem under discussion. I think that if you take a look at what's going on at the moment, you can get an idea of the desire of certain members of Congress for some good advice. Congressman Mel Price has just been named chairman of a new subcommittee on research on the Armed Services Committee, which gives him a chance to take a crack at about $7 billion in Defense Department research funds. And he has a first-class group of people working with him. We already have the Daddario and the Elliott committees. So you see, the proliferation of effort continues.

I would suppose that one way Congress could attempt to deal with this problem is by building up its committee staffs, because the different staff directors of the various committees are certainly going to be loyal to the congressman who saw to their appointment, and at the same time they would be informed and inclined to communicate with each other.

Comment: One of the difficulties, though, is the traditional reactions of Congress to a problem—namely, to investigate what is wrong, to put quick labels on something, or to white-wash the problem. So that the upshot of the whole thing may be a committee to investigate why so much money was spent on a study of such foolish things as life among the apes.

Mr. Brademas: Let's not prejudge the point. We don't know what these committees are going to do yet. So far, I think you would agree, the Elliott committee has conducted its hearings in a very open-minded, fair manner without any particular ax to grind.

Question: As a hard scientist, I'm terribly interested in the impact of science on society as a whole. I represent a large group which has members on 250 campuses all over the country, and I've been wondering whether, without any remuneration and without being appointed to your staff, one of us could offer his expertise in a very specific field to the members of Congress. Is this feasible? Is this practical? Or would you in Congress be inclined to refuse our offer?

Mr. Brademas: I think your suggestion is a splendid one. I'm sure that problems would arise, but I think the benefits would be very great indeed. Obviously not all members of Congress would be interested in this kind of an arrangement, but there would be enough of them so that you could make a most valuable contribution. I'd be glad to forward this suggestion or invite you to come to Washington sometime and discuss the idea with some like-minded colleagues of mine.

Comment: I appreciate your geographical loyalty but I would like to suggest that in discussing these matters one should view them from a larger perspective, from the viewpoint of the nation as a whole.

Mr. Brademas: While we in Congress are representatives of specific districts, we're also United States representatives. And I must say that I vote for a lot of programs that have absolutely no impact on my district.

Comment: I was deeply troubled by your description of

the scramble for defense money. Now, if defense money is used to masquerade sound investment in science, I'm all for it. But I think that most of the time science is confused with mere technology. I wonder if you in Congress are not in a difficult position to decide what is science and what is mere technology in the interests of a certain group.

Mr. Brademas: I think that may be a fair point, but so far as the geographical impact is concerned, members of Congress are not too troubled over whether the money is labeled science or whether it's labeled technology.

Question: I am somewhat puzzled by your reference to the need for expertise. I would have thought that the real problem was the first one you mentioned—time. It seems to me that many of the choices you in Congress are confronted with are not so much hard to understand as they are hard to decide. That is, if you had enough time to listen to all the alternatives, you would be able to make fairly satisfactory decisions; that it's not so much your lack of technical expertise that handicaps you, it's finding the time to consider the problem in detail.

In this connection, I would like to ask you, to what extent is the experience of the Joint Committee on Atomic Energy typical and relevant? Here is a committee that over the years has apparently been fairly active in influencing executive policy in a very technical field; yet its members have had no built-in expertise, no scientific capability. But they have had power. And to the extent that they have had power, they have been able to rely on the experts in the executive branch to supply them with whatever information they might need about programs which one person or another felt to be deficient or in need of change.

Mr. Brademas: I'm not familiar enough with that committee to comment. I do know that it has become powerful, and I've operated on the assumption that among the reasons it has are these: first, there are able men, both senators and congressmen, sitting on the committee; and, secondly, its members have built up an able staff—so able that, combined, it can challenge the assumptions of the executive branch. And this capacity isn't something they've achieved overnight.

Comment: They've also had the authority to inquire into these programs, which the other committees don't.

Comment: Still, they don't have the technical expertise that you pointed to earlier as being so essential.

Mr. Brademas: I don't know. I thought they did have some able technical, scientific people on their staff.

Comment: Who is the expert in this sort of thing? I think the whole federally supported scientific effort is much too new for us to be sure that a lot of people who are now considered experts are in fact experts. For example, of the many persons who have testified before the Elliott committee, a majority have been physical or natural scientists. The expert in these areas has experience which gives him some knowledge and insight about a lot of these problems. But it's a matter of his experience and not in any sense a matter of his technical know-how. There's no reason in the world why a physicist, however good he is, is an expert on what the nation should do about biology or even physics.

Mr. Brademas: I think I have too narrowly made my point. Where I used the term technical expertise, you can substitute, if you will, experience, judgment, or knowledge. Let me give you an example from my own experience. I spoke of having an engineering dean on my staff. I was interested in having him, not because he knew a lot about heat engineering but because he was an engineer; he understood the jargon of engineers. He was an educator; and, with respect to a particular bill of mine, a technical education bill to encourage the training of more two-year semiprofessional technicians, I wanted somebody who understood the subject-matter and was adaptable and intelligent and able to call on the lobbyists whose support I wanted to generate.

I couldn't detail a member of my staff to this task full time. First, he wouldn't have had the time, and second, he wouldn't have had that kind of judgment and experience. It's assistance of this sort which I think would be useful to me.

Comment: All I want to point out is that we ought to recognize that we don't really have experts in the sense that we usually use the word "expert" because the field itself is just too new.

Question: On the point of what kind and what quality of scientific advice is available to Congress, the National Academy of Sciences and the National Research Council by their charters are supposed to provide scientific advice to the federal government. What is the attitude of members of Congress toward these organizations?

Mr. Brademas: In the first place, I think most members of Congress never find it within their experience to have any dealings whatsoever with the National Academy of Sciences. Many members of Congress aren't quite sure what it is. The kind of a discussion we're having just wouldn't go on in the House of Representatives. This isn't the sort of problem that a member of Congress ever considers because he just assumes that the National Academy of Sciences has nothing to do with his problem. Now, if the National Academy of Sciences were to undertake what you have suggested, to come to us and say, "We're interested in seeing to it that members of Congress have access whenever they need it to scientific expertise," that would be a different matter. But so far they've not done so.

Let me make another point which your question brings to mind. Take the National Aeronautics and Space Administration, for example. It's only since it began getting into trouble with Congress that NASA suddenly began to say, "Good afternoon, Congressman. How are you today?" i.e., to give a little attention to congressional liaison. So there has to be some built-in reason for contact between Congress and the National Academy of Sciences, just as there is now between Congress and the NASA people. It isn't going to happen just because congressmen suddenly decide they should learn more about science.

THE SOCIAL SCIENCES

Question: You have been talking about science with a capital "S," meaning the physical sciences to a certain degree. But what about sociology, psychology, or political science? Aren't these other disciplines important in coping with the problems being created by some of the hard sciences?

Mr. Brademas: I think that when most members of Congress use the word "science," they are thinking about the natural sciences—the hard sciences, if you will. And there can be no question that there is a lot of hostility toward providing federal support for research in areas like psychology or sociology. The very mention of these fields is likely to occasion a bitterly hostile reaction from some members of Congress, as if they are obviously subversive disciplines. When the American Orthopsychiatric Organization recently came to Washington, some of us talked with its members and found they had the feeling they were neglected by Congress. One of the reasons I suppose is that they've never made any great effort to make themselves understood to members of Congress. But I really think it's a deeper problem that that. One can, I think, easily justify in terms of national security the investment in the hard sciences. It's much more difficult to justify a vote for research in sociology or social psychology to the people back home. Moreover, these fields are likely to be more controversial since they involve points of view about society.

Comment: Could I just say in passing that I think you did yourself a disservice? The Congress has over the years moved a fair distance toward providing research money for the social sciences. Just very quickly I could cite the Children's Bureau, which is now getting more money; the Welfare people, who are finally getting some money for research; the manpower people in the Labor Department, who are getting research money under Title I of the Manpower Act; the social science division of NSF; the Department of Defense, which has always provided a fair amount of support for social science research; and, more particularly, the National Institutes of Mental Health, which have been the major subsidizers of social psychology and psychology.

Although I think there are problems here, I wouldn't mini- mize the amount of money that has been appropriated by Con- gress for research in the social sciences. Most of the money has come by indirection, it is true, but that may have been the only way it could have been provided.

Question: I was impressed by your remarks about the po- litical aspect of your job and the juggling that has to go on. I was also impressed by your comment about the failure of the so- called social scientist to communicate in a way that is meaning- ful. I wonder if you have any thoughts on what those of us who are social scientists can do to lend our expertise in a more articu- late way to help you.

Mr. Brademas: For several years now, the American Politi- cal Science Association has been sending around interns to serve on the staffs of members of Congress for three months. There are two kinds of people the APSA send—social scientists and journalists. After a little experience, I have all but written off the social scientists. They just don't help in the operative, pragmatic, practical, day-to-day work of a congressional office. There are a number of ways, however, in which social scientists could help us. For example, I could have used some social scientists—say University of Michigan survey-research type people—to go into my congressional district and see whether the Studebaker work- ers blamed the shutdown of their plant on me or on the manage- ment. It is extraordinary to me that the academic social scientists have not sought to make available their talents and knowledge to

a greater degree then they have to the members of Congress.
There are all kinds of things social scientists could do to
help us in our job. You may define these as partisan, of course,
but that's the business we're in.

EDUCATION

Question: I have a question which is addressed not only to
Congressman Brademas but really to all the members of the
seminar. It seems to me that we've concentrated here on what
the scientist and technologist can do to advise the federal government. I wonder what impact these bills for federal support of
education for science and technology will have on science and
technology. Much of the federal money is for defense and space
research and this is bound to affect the science and technology
of the future. A good many of us have had experience in this.
What we do is practice a little deception in order to tie up our
research interest with defense or space. This isn't completely
honest on our part, and it worries me as an educator with certain
moral responsibilities.

What does this linkage to utilitarian space and defense applications mean in terms of the direction science and technology
will take in the future? And what direction will education take?
I can see this in historical terms as an extension of the federal
government's responsibility to aid education, and I'm all in favor
of that; but at the same time I know what happened to agriculture when the federal government began sponsoring research in
that area. Agriculture has really been the most amazing phenomenon in American science and technology in so far as increasing productivity and decreasing the labor force is concerned. If science and technology are going to have the same
impact on industry, we're going to have a population that just
sits back and watches while the machines do all the work. This
suggests enormous social, political, and economic implications.

Mr. Brademas: You've posed a range of questions, each of

which is pretty profound. I think one of the previous commentators was not as troubled by the moral problems as you are. He said that it was perfectly all right with him if we could get some money for education and science under the guise of defense.

Let me just touch on one of the problems of which you spoke, and that's the impact of putting all this money into only certain selected areas. Mrs. Green, the distinguished and extremely able chairman of my subcommittee, recently released a report on the federal government and education, with which some of you may be familiar. She was deeply concerned that we're spending so much money on the hard sciences and so little on the humanities—and so am I. But frankly it was all some of us could do to persuade her not to include in that report a recommendation that we slash our funds in the hard sciences and spend more money on the humanities. Our argument was that we should put more money into the humanities. There are many good reasons why we spend more money on science than we do on the humanities, and they are not only ideological. The high cost of scientific equipment is only one. Congressman Daddario's report takes note of this problem of federal spending on science as compared to the humanities, and it is a recurring theme in congressional hearings on science or education.

Question: I wonder whether you would comment a little more on the vocational education bill because I gather you not only restated the priorities, you added categories and perhaps even developed a whole new philosophy of vocational education.

Mr. Brademas: Yes, we did. Here are some of the things in that bill. We provided a lot more money. We authorized the transfer of funds from one category of vocational education to another by the state vocational education authorities. We didn't make it mandatory that they do so, but we did make it possible for them to take cognizance of the fact that people have moved off the farm into urban areas, that people must be trained for office as well as farm occupations. We also provided training in farm related operations, such as the processing of foods. We required that vocational educators talk to the employment security people in their states so they can get some idea of what the local manpower needs are and gear their courses to them. We established a new work-study program which is somewhat controversial. Young people who otherwise wouldn't have the money can obtain a type of scholarship to go to school and get same vocational training—they can work and also study. We authorized the construction of some area residential vocational schools. Fundamentally speaking, this bill is an effort on the part of Congress to take note of the urbanization that has taken place in the United States since the first vocational education law was enacted.

Question: In the vocational education bill, was there any consideration given to the segregated vocational education schools in the South?

Mr. Brademas: We built no civil rights rider into the bill. We hope the same end will be accomplished by administrative pressure.

Question: Is there any prospect that that will come about?

Mr. Brademas: I really don't have enough information to give you. I know that the NAACP came in and made a very strong plea for a civil rights rider. I have to add one other point. This vocational education bill is a civil rights bill, too. I think you all understand why.

SOUTH BEND

Question: What so far, if anything, has the Federal Employment Service done about the Studebaker situation? What steps have you taken to get action?

Mr. Brademas: A representative from the Bureau of Employment Security, an expert in dealing with the problems of training workers who have been unemployed in a large plant that is suddenly shut down, came to South Bend. One of the government's top labor market analysts also came to South Bend, because one of the morale problems in South Bend was rumor, with nobody quite sure how many people would be thrown out of work. So we had to get the facts. This was important, not only to dispel some of the rumors flying about, but also to determine whether South Bend was eligible for classification under the Area Redevelopment Act to receive accelerated public works assistance. But the community leaders were very sensitive about being labeled a depressed area. They were not quite sure if they wanted to be so designated. I had to find out how they felt and explain to them that they couldn't have it both ways. We got in touch with the Federal Housing Commission to discuss the problem of workers who were going to be unemployed and who had mortgages on their homes. We arranged that whenever an unemployed worker went to his lender and asked for forebearance on his FHA loan, the FHA would, if the lender agreed, automatically approve that arrangement.

We got the Secretary of Labor to send a letter to all companies in the Midwest who we had reason to think would be interested in recruiting workers from the Studebaker plant. The Secretary requested that these recruiting efforts be handled through the Indiana Employment Security Office in order to maximize its efficiency. In South Bend our motels and hotels were filled with Chrysler, Ford, and other recruiters who came to hire former Studebaker workers. Under a contract with the Area Redevelopment Administration, a first-class man in working on this kind of unemployment problem, Dr. Harold L. Sheppard, of the Upjohn Institute of Employment Research, came to South Bend to coordinate federal activities on the scene and see that all resources are made available.[4] We got the UAW and Labor Department to agree that they would speed up manpower training programs. South Bend was one of the first cities in the country to have projects under the Manpower Act, so we were ready to do business.

Efforts were already under way to organize space for the mass registration of unemployed workers at the UAW hall in South Bend so that they could make their claims for unemployment compensation from the state of Indiana. At the same time the Labor Department sent in extra counselors and interviewers to sit down and talk with these unemployed workers about other job opportunities in the area.

I also was in touch with the Bendix people, because they are a very large employer in the South Bend area. This was to see if Bendix could absorb any of the people. Three hundred people used to sound like a lot of people to be thrown out of work, but when you have 6,000 people out of work, then if you absorb 300 of them here and 600 of them here, it's all to the good.

It was fascinating to see all those top federal officials sitting in a conference, and you know, they really cared about what went on in South Bend. A key problem was the government contracts that Studebaker had. We hoped that Studebaker would perform those contracts in South Bend, or, if the facility were to be purchased by another firm, that they would be performed in South Bend.

At the same time, we suddenly went into the industrial development business. The Indiana senators and I were on the phone around the country, in touch with major corporations who might be prospective buyers of the Studebaker facilities.

Question: Did Studebaker give any advance warning to

[4] Dr. Sheppard's summary of the actions taken following the Studebaker shutdown, with his recommendations for appropriate steps in similar situations, has been published as a case book by the Area Redevelopment Administration (1964).

you or to the Labor Department that this was going to come about?

Mr. Brademas: This is a rather sensitive matter. The answer is "no." In Germany I don't think a company would be permitted to shut down a plant of this size.

Comment: I don't think there is anywhere a company could get away with it except in the United States.

Question: Has it occurred to anyone to lower the taxes for five years so that an outsider would become interested in moving in?

Mr. Brademas: That's a very good suggestion. I was told by the Indiana senators that there were some pretty good nibbles for the Studebaker facilities and tax abatement may not be necessary, but if it takes that to do it. . . .

Comment: It seems to me that the response of those federal agencies to the situation in South Bend is more or less dependent on the imagination and will of the officials concerned to slightly distort the function of their agencies to meet the situation. And strictly, after the fact, they might be criticized for doing exactly that.

Mr. Brademas: It may be true that they had to depart somewhat from their normal functions, but they had other situations as precedents, such as when the Packard plant shut down, or when the meat-packing plant in Omaha closed its doors. I noticed that in our conversations they would often say, "Well, this was the way we did it in Philadelphia," or, "This was the way we did it somewhere else." I must say that I was impressed, sitting around that table for a couple of mornings, by the high quality and imagination of the bureaucrats. Of course, these are top bureaucrats we're talking about.

Question: The thrust of the question is: Shouldn't there be some institutional provision for such situations?

Mr. Brademas: That's a very fair point and a telling one, it seems to me.

Comment: Senator Clark held hearings on the manpower revolution. After six months he had taken seven volumes of testimony. The Senator had some foresense that defense spending was going to level off and perhaps even decline, and he has been bothered about this point, that is, how do you institutionalize forward planning. Of course, the Department of Defense now has an Economic Readjustment Office. But I think Senator Clark feels that the government hasn't yet developed adequate structures to cope with this problem.

Question: Could you tell us something about the demographic characteristics of the workers displaced in South Bend?

Mr. Brademas: This was one of the real problems. The average age of the worker in the Studebaker plant was about fifty-five. He thought of himself as an automotive worker, and he thought of himself as a Studebaker worker, because there was very intense loyalty to that company in my town. He hadn't looked for a job for a long, long time; and the very idea that for the first time in, say, twenty-five years some of these men were actually going to have to go out and look for a job was, in my judgment, a very severe emotional and psychological wrench. I was also told by people who deal with these problems that it would be at least two to three months before the workers in South Bend who were unemployed would actually realize the situation they were in.

Question: What was their average earnings?

Mr. Brademas: At least $100 a week. They were home owners, which created another problem. Since most of them owned their own homes, they had a financial as well as an emotional stake in the community. I daresay that those who took jobs elsewhere would probably want to come back if something were to happen so that the Studebaker plant reopened again.

Many of those displaced were probably semiskilled. They needed some retraining if they were to find employment in such fields as, say, electronics—if such an opportunity were offered.

So from the point of view of the older Studebaker worker, it was a rather sad picture.[5]

Question: I'd like to ask two questions about South Bend. One, what is the present rate of unemployment in the city? Two, what percentage of the labor force did the 6,000 Studebaker workers represent?

Mr. Brademas: The percentage of unemployment before the announcement of the Studebaker plant's closing was 2.1 percent, which was the lowest in ten years. When President Kennedy was inaugurated, the time from which I date these matters, my home county had an unemployment rate of 11 percent. You can thus see the tremendous improvement we've made. There are about 93,500 people in that labor market area; so the Studebaker workers represented about 7 percent of the labor force.

Question: You spoke with some admiration of the bureaucrats from the Federal Employment Service coming into your area. But I know that one of your colleagues from Ohio, among others, has been attacking the Employment Service. How successful do you think he'll be in emasculating the Employment Service?

Mr. Brademas: I don't think he'll be too successful because it's fundamentally a popular, widely supported service.

RESEARCH AND DEVELOPMENT AND ECONOMIC GROWTH

Comment: As you have indicated, these industrial corporations which obtain developmental contracts tend to get production contracts as well. But I doubt whether the corollary that basic research in the universities or similar institutions tends to attract industry to that area, is also true. So maybe before you break your back trying to get the federal government to invest funds in pure science research in South Bend, you should make

[5] In 1964 South Bend received funds under the Manpower Development and Training Act for the establishment of a demonstration project, Project ABLE (Ability Based on Long Experience) aimed at helping meet the needs of older (fifty years and over) unemployed workers.

sure that it will have the economic impact in your district that you claim it will.

Mr. Brademas: I agree with you in substance, but speaking now as a politician, I couldn't disagree with you more. There are two reasons for bringing a research center to South Bend, Indiana: One is that in the long run it will generate more economic activity, though perhaps not as much as what I said earlier might lead you to believe. The other reason was given by one of my friends back home when with great fervor he urged me to get busy on the NASA center. "Let's get one of those brick and mortar things in here," he said, "something they can see." Politically speaking, completely removed from the ecomomic consideration, there is still going to be a scramble for these installations.

Comment: The fact is we don't know exactly what the effect of scientific development is upon economic and social development. This is something which calls for further research. We don't know, for example, the regional consequences. And maybe the consequences are not regional, but institutional.

Question: You have pointed out that most of the federal research-and-development money goes for national defense. Is there any committee working with the research-and-development people in the Pentagon to see what kind of inventions that might produce jobs are simply lying around, perhaps because of some nonsensical restrictions? In other words, is anybody considering whether all these things which are dreamed up for reasons of defense might have some civilian application?

Mr. Brademas: There seems on the record to be very little fallout for the civilian sector of the economy from federally supported research and development in space, defense, and atomic energy. For that reason we need to make a head-on attack on the problem. Now, there is a section in NASA, I think called Civilian Utilization, or something like that. But I think it's primarily window dressing to keep irritable congressmen off the backs of NASA.

Question: Isn't there some way so that when a situation like South Bend develops some of these new products lying around in files can be pulled out and used to produce jobs?

Mr. Brademas: I don't think the economy works that way. My own feeling is that if somebody had such new products, he would not keep them lying around.

Comment: Maybe Mr. Brademas' neighbor, Dr. Edwin Gee, who runs the development department at DuPont would comment on that. He's been at both ends, and that's what he's supposed to find out: is anything lying around?

Mr. Edwin Gee: No, I don't think technological innovations are just lying around waiting to be exploited. A lot of people are trying to milk the research-and-development effort day in and day out. A lot of money is being spent both by industry and by the government. But people aren't finding as much as one might think. And I think that the reason is pretty straightforward. I was a chemist once, and there's a very large number of compounds in any area. But a count of how many of them people use shows a very, very small number. This simply means that in the total output of science, there's only a certain amount that is useful. I think the process goes like this: First someone must make a scientific discovery; then an inventor has to come along and put it to use. But there's a third step in the process, one which doesn't occur in most situations. An entrepreneur has to come along to figure out how to provide the momentum to make the product, to sell it, to make a profit, and to keep it going. The kind of government research, in which I spent ten years, is not directed to making a profit. So when it comes up against the enterprise system, where it's necessary to make and sell something, it's pretty poor at it—because this isn't what it was intended to do.

Question: But can it be transferred? The jet engine, which was a military, not a civilian invention, eventually had a civilian application. The same was true of radar.

Mr. Gee: Yes, that was true, but only over an extended period of time and with an importance that was very small in relation to its importance to the military. Sure, these things do happen. And the example in Mr. Hollomon's files that stands out best is the use of Teflon [6] by the military. In tracing its origin, it turns out to have been a chance discovery in the laboratory of an industrial concern, and because it had certain properties, it was used by the military. I'm sure there are other examples. But they're chance results, not planned objectives.

[6] A plastic which has the property that it is nonadhesive.

Modern Science and Its Implications for the University

by RALPH S. HALFORD

Dean of Graduate Faculties, Columbia University

THE AMOUNT of federal funds spent in support of research at Columbia University may approach one-half of the university's total expenditures from all sources during the current year. During the past twelve years the volume of this support has increased more than tenfold.

What has been the effect of such a massive injection of funds exclusively for research upon the affairs of the university? The dosage has reached proportions where its consequences should be no longer a matter for conjecture. Is the university being transformed into a giant research institute?

My answer is emphatically "no." Students are not being crowded out by a preoccupation with research. We have in most schools about the same number of students today as ten years ago. The number of their teachers has increased somewhat, but mostly, it seems, in fields that do not receive government support for their research. More courses are being offered today than ten years ago; so the students are enjoying if anything richer fare. There is no evidence of any general harm to the instructional program, and it would appear on balance to have been improved.

If these observations seem unexpected, they have a simple explanation. Teachers at Columbia have always been free to devote about one-half of their time to scholarly research. They can spend this time more fruitfully when their research receives generous financial support as it does today. Moreover, the researchers' performance during the portion of their time devoted to teaching is a reflection of the quality, variety, and timeliness of the knowledge they are able to impart to their classes. The closer one is to the source of new knowledge, the more quickly he can learn of it.

To prevent undesirable consequences from massive support for research is a matter for management, and I submit that Columbia has managed well. We will solicit support for research projects initiated only by persons who are employed here primarily as teachers. The university does not appoint persons to its faculties to devote themselves exclusively to research. Professorial titles signify that the holders are expected to teach regularly. In this way the number of persons who can enjoy support obtained from the government with our auspices is governed by the requirements for their services in our instructional programs.

Important changes can be discerned otherwise, of course, in the pattern of the university. We have acquired a large supporting staff to expedite the search for new knowledge. Their presence creates new demands upon our space. Here again, however, by restricting any encroachments upon space required by our instructional programs, these can be protected from harm. Meanwhile, the remaining portions of our physical plant are utilized more fully, and for the purposes they were intended to serve. The change is for the better.

So much for the past and present. What about the future? Much has been said, and often in alarm, about our universities' growing state of dependence upon the federal government. I would like to suggest that this proposition be examined the other way around. Society is dependent today as never before upon our great universities. As primary sources of new knowledge and communicators of the old, they both lead the way and supply the followers from those other institutions that stand ready to apply the knowledge to serve the public's needs. There are fewer than fifty universities in this country that can respond to this dual challenge in any significant way and less than half of them can do so both broadly and importantly. Among these latter, Columbia is at the very forefront. For us to strike a posture of independence would be to stand aloof from society's needs. Society would despise us surely in that pose and would turn its back upon us totally. We should be proud that what we have taken always to be our self-appointed task, the search for new knowledge, has come to be of such importance to national health, prosperity, and security that society now chooses in its own self-interest to encourage us generously in that task with public funds.

In short, if we have entered recently into a new relationship of dependence, that dependence is mutual and the burden of the greater need is on the other side. We have managed our affairs so that we can resume our former posture undaunted if we should be cut off from public funds at any time. It is doubtful whether our society could prosper for long if deprived of what we now furnish it with those funds. The relationship that has developed between Columbia University and the federal government is neither compromising nor precarious for us and, barring loss of faith in the power and worth of new knowledge, it can and will in its present fashion endure.

This new relationship is not a parochial one. We deal with a number of agencies of government and with many autonomously funded programs within each of them. Their support flows to more than 750 projects here, concentrated in the sciences, to be sure, but placed in total throughout the length and breadth of the university. Expressed roughly, about 40 percent of the funds they furnish goes to our School of Medicine, another 40 percent to the Faculty of Pure Science, and 11 percent to the School of Engineering and Applied Science. But the remaining 9 percent represents a substantial sum in its own right, on the order of $3,800,000 annually, and portions of it are directed to activities associated with the School of Business, the School of International Affairs, the School of Journalism, the School of Law, the School of Library Science, the Faculty of Political Science, the Faculty of Philosophy, and the School of Social Work. These activities alone receive about as much support today as the sciences did twelve years ago. We must bear in mind, however, when making comparisons of this sort, that the paraphernalia of research are not similarly expensive from field to field, that one thousand dollars may furnish as much implementation of scholarship in one field as ten thousand in another, and perhaps even as much as a million dollars in a third.

The scholarly imagination, once it has been invited and assisted to soar, can be extremely expensive to maintain in its free flight. There is serious discussion in progress, both within government circles and among scientists at a number of universities, concerning a proposal for a device that would cost an estimated one billion dollars to build and perhaps one-tenth that much annually thereafter to operate. It would serve the immediate purposes of not more than a few hundred, if that many, nuclear physicists. Yet it will be indispensable if the quest for knowledge in this area is to continue.

Modern developments in science have established dramatically in the public mind the worth of new knowledge. They have brought in their wake vast public support to implement the search for this new knowledge and to develop its useful applications. This support has ramified into areas remote from science to encompass today every kind of knowledge that might be utilized somehow to promote the public welfare. All aspects of the university have been strengthened through the resulting injections of public funds, indirectly if not directly, and none have been weakened.

The university might appear to some to have been greatly

changed, but it has not altered its intended aims or even shifted its own internal emphases upon them. The implementation that is being newly furnished to the search for knowledge permits us, at least in some areas, to accomplish that one purpose fully for the first time. We speak to the world more forcefully today, and it listens more attentively, than ever before. If our accents seem to have become misplaced, it is only because some of them need still to be strengthened, not because the others have become too strong. We should seek to restore the balance of our former diction, not by quieting what has become newly powerful but by reinforcing our other efforts in the search for those kinds of knowledge that find their applications for the benefit of society in less tangible ways, through the spiritual betterment of the individual man.

As of December 31, 1963, there were 748 actual funding agreements at Columbia providing a total of $44,256,800 in funds on an annual basis. Now, my reason for presenting these figures to you is to try to give me a little authority for the following statement: $44 million in university circles, and I think in probably any circle, is big money. It invites the easy but erroneous conclusion that we are dealing in big science. We are not. This is the first point I would like to make.

Nearly 750 proposals at a total value of a little over $44 million is something less than $60,000 per funding agreement. If I may be permitted to reduce the total of the funds by removing the ten largest projects, then we have perhaps 740 agreements at something under $30 million. The average amount of funds per funding agreement is in the order of $40,000. Actually, the median is quite a bit less than that.

This is still a very handsome sum of money. My colleagues in the social sciences or the humanities would, I am sure, consider that a very handsome sum, indeed. But what do you buy in academic research today with $40,000?

About 40 to 50 percent of it goes into salaries. This average project then would pay something like $16,000 to $20,000 in salaries. Who gets these salaries? Under this kind of program graduate students in the sciences are subsidized, receiving from $3,000 to $4,000 a year each as nominal half-time employees. Post-doctoral education, although it is not recognized by degrees or certificates, has become a very important part of the educational process, and the post-doctoral research associates will receive anywhere from $6,000 to $8,000 a year. Two graduate students plus two post-doctoral associates is about $20,000. That's a typical project, or four graduate students and one post-doctoral fellow.

About one-sixth of the money in the project will go into indirect cost allowance, or, as others refer to it, overhead. I prefer the more euphemistic designation. The balance of the money, representing maybe a third, goes into the expendables, the materials, the supplies, the equipment that are used by the scientists in the course of their research. I haven't wide familiarity with this, but I know that for every graduate student in chemistry (the department where I have spent most of my life at Columbia), in one way or another, whether it be federal funds, or university funds from endowments or otherwise, we spend on the average more than $1,000 a year for expendable supplies, solvents and other things that go down the drain when the experiments are completed. Scientific research is expensive, but this still is not big science. We are dealing here with little groups, little cells of four or five people whose activities have to be supported from day to day.

I'd like to turn to another point. In reading the proceedings of this seminar, I have noticed a tendency to deplore, or, if not to deplore, at least to comment on the fact that most scientific effort these days seems to be involved with the Department of Defense and the Space Agency. It would be a false generalization to assume that this describes a university effort. Of the $44 million in federal funds expended at Columbia, $16.5 million comes from the Public Health Service, $9 million (about one-fifth) comes from the Navy, $3 million from the Air Force, and $1 million from the Army. About $4.7 million comes from the Atomic Energy Commission, and $3.9 million from the National Science Foundation. Only $800,000 in upwards of $44 million (less than 2 percent) comes from the Space Agency. This sum is almost matched by $692,000 from the Peace Corps. In other words, the university is engaged very heavily in basic research, while on the other hand the larger proportions of the money appropriated by the federal government go for development and applied research. I want to make it clear that the university has not abandoned its principal mission, that of seeking new knowledge, for the mission which is much better left to private industry and other organizations, that of applying this knowledge to technology.

These sums represent a large amount of money in absolute terms and also a large portion of the total university budget. But those of us who are concerned with the administration of these funds do not try to direct what is done. Our oversight is strictly permissive, seeking to protect the university against eventualities, some of the hypothetical, but to us very important.

What sort of safeguards do we have against what you might call the total catastrophe in the hypothetical event of an abrupt withdrawal of government support? This would be a catastrophe, and I don't want to minimize it. It would be a catastrophe for science, it would be a catastrophe for the nation, it would be a catastrophe for the academic profession. But what I would like to emphasize is that it would have relatively little effect upon what I will call the integrity of this institution. We would be fully prepared to honor all the tenure commitments and all the other necessary commitments for teaching services out of our ordinary funds. We do not use government funds to relieve the university of its salary obligations to teaching personnel. The only exceptions to this are those occasions, relatively rare, when individuals are relieved from part of their teaching duties in return for a reimbursement to the university from government funds, which are then used to employ a substitute teacher. In the event of catastrophe we could dispense with the substitute teachers and return to the *status quo ante.*

Then, too, the university has a certain momentary dependence upon the indirect cost allowance. I will elaborate on this if you wish later. I will simply say now that the amount of the loss from indirect cost allowance which could not be immediately compensated for by a cessation of expenditures is roughly equivalent to the increase in the university budget that normally takes place in one year, so that in the event of an abrupt and total cessation of government support we would be in a bit of financial distress for about twelve months, and then I expect that everything would get back in balance and there would be no particular problem.

A much more subtle thing would be distortions. What safeguards do we have against those? This is arguable. Distortions could accrete little by little, and gradually the university would become a very different place. No person in my capacity or any other in the university can possibly comprehend what is the substantive content of 750 different kinds of research activity, branching all the way from the field of structural linguistics to recondite areas of medicine. Still, we have to be concerned to try to avoid initiating inappropriate activity. This is not as difficult as it seems in a mature program such as the one we have at Columbia, because most of our government funding is for renewal or continuance of something that has been done before. There are relatively few new things being introduced, and so we can employ our watchdog procedures to look primarily at new proposals. We have to be very careful also about the avoidance of what are possibly appropriate but completely unmanageable activities.

It would be totally out of the question for this university, in my opinion, to undertake the management of the contemplated billion-dollar accelerator device with its operating budget of one hundred million dollars a year. This would be a bigger thing than we are involved in now, and we would become something totally different for doing it.

We and a group of other universities on the East Coast are involved with an organization, Associated Universities Incorporated. Our involvment is something that I leave to the economists and lawyers to describe technically, but Associated Universities is a nonprofit corporation formed for purposes of education in the state of New York. Among the twenty-five trustees allowed by law, each of nine universities selects two. One must be a scientist, the other an administrator in his institution. There is nothing to bar the latter from having had a scientific background like myself, and so most of the trustees are, in fact, scientists. This organization acts as a buffer between the universities and the government. It holds the contract from the Atomic Energy Commission to operate the Brookhaven National Laboratory. More recently it has acquired a contract from the National Science Foundation to operate a national radio astronomy observatory. It specializes in projects involving enormous facilities and large staffs. By and large its function is to insulate our universities from the undesirable effects of "big science." It, and not Columbia University, is the appropriate agency for managing "big science."

Finally, as still another safeguard against distortion, there is the firm, the devoted insistence, upon the recovery of all indirect costs related to the performance of research under government sponsorship. In order to prevent hidden and inequitable subsidies to the sciences out of our precious university funds, thus starving the humanities or the social sciences, we are adamant upon this point. Our colleagues do not always understand this attitude, but we insist that we must recover all of the indirect costs, and every university in the country feels similarly. It is not a simple matter. The determination of indirect cost is in the first instance problematic, and in the second instance mighty tempting. When Professor X comes in to discuss a grant of $40,000 that is $500 short on indirect costs, there is a great temptation to say that this does not amount to very much and to approve the project. But when this sum is multiplied by about 750 projects, it becomes real money. And so university administrators have to be tough in each individual case. All I can say of our performance in this area is that, given the government's own audit, not our calculations, of what the university's allowable indirect costs are, we are recovering at the present time only about 80 percent of them, because of limitations which Congress imposes on reimbursement for indirect costs in appropriations bills for the Public Health Service, the National Science Foundation, and certain other agencies.

Finally, there is the very obvious problem of avoiding any creeping expansion based entirely on government funds, and I think I have indicated how we at Columbia have tried to avoid this. Our principle is to relate our requirements for faculty to the instructional requirements of our teaching programs. We use university money to bring to the university the people we need to instruct our students. I don't mean this in a narrow sense, for I also mean instruction of graduate students in research. It is only for these teaching personnel that we will seek assistance from outside sources to enable them to do what they want to do in the way of searching for new knowledge.

In recent months there has been considerable concern about the fact that federal grants suddenly seem to have reached a sort of momentary plateau. I don't know whether this is the proper time for it, but it does not seem surprising to me at all, for I think that the possibility of saturation has been with us for some time. With a fixed number of people, if we do not expand the university staff in order to use government funds, there must come a level of support at which these people's activities have been as fully implemented financially as possible, and beyond this point it would become inefficient economically or otherwise to furnish them with more funds. I leave it to the social scientists to decide whether the wisdom of Congress has determined correctly that this level has now been reached, but my own way of deciding this from the mass of statistics which is prepared for my office is to examine the monthly expenditure reports to see whether we have been spending the funds that are available, And it is only within the past six months that I have detected clearly and unmistakably that we at Columbia are underspending some of the funds that have been committed to us. I don't mean that this is true across the board (it is, in fact, true only in certain highly localized areas) but it is a significant and unmistakable phenomenon. It is interesting, or perhaps just coincidental, that in light of this fact the collective wisdom of the political process had decided that the time has arrived to question the amount of money that is going into research.

DISCUSSION

THE INTEGRITY OF THE UNIVERSITY

Question: What do you mean by the "integrity" of the university? The fact that Columbia University can financially support its commitments to its tenure faculty strikes me as nothing more than financial integrity.

Mr. Halford: I mean integrity in a historical sense. The university was something in terms of a financial pattern, something in terms of a set of principles in 1939, when the phenomenon of large-scale government grants was unknown. The university has protected as best it can what it was, what it stood for at that time. We could return to what we were in 1939.

Question: What would the university look like if it went back to 1939 and this meant a loss of one half of its income? What would it mean if you lost those members of the faculty who have at least part of their salaries paid out of the research funds?

Mr. Halford: People could go back to doing research as it was carried on in 1939. It would not be a serious problem. A year ago only about one percent of the university's tenure personnel were being paid by government funds.

Question: $15 to $18 million of the federal funds expended at Columbia is from the Public Health Service. Most of that goes, I gather, to the medical school rather than to the life sciences at Columbia University. At one time medical faculties stood in relation to the life science faculties, biology and zoology, much as engineering faculties stood to the pure science faculties. Hasn't there been an extraordinary shift in balance here? Is it possible that basic research in the life sciences perhaps is not as well supported as the figures would suggest? This doesn't apply just to Columbia, but to any university which has a medical faculty as well as a biological faculty.

Mr. Halford: If there is an imbalance in available funds between medicine and biology this is not something that one can control by local administrative action. This arises from the fact that Congress thought that more money in medicine would be desirable. It is true that if we use the 1939 date the medical school is a much more expanded activity than it was, but in terms of the university's obligations to its tenure staff it is no more expanded than our biology department.

Comment: The large grants to medical schools are partly a way of supporting post-graduate medical education. There are no funds available otherwise, and the need for such post-graduate medical education has become acute during the past few years.

The school of engineering today is doing a lot of work that was considered to be in the realm of physics twenty years ago, and I am sure that the medical school does work which only twenty years ago would have been considered biology.

Comment: All disciplines compete for students. If all the rewards, or the large rewards, are in science, the best people will be drawn into science. And therefore society will become scientific. I disagree therefore with those who say the fact that Columbia is now scientifically oriented is not changing the nature of the university. I think the issue is whether or not it is a good thing.

Comment: A most interesting phenomenon is that over the past fifty years there has been no percentage change so far as we can determine between those of exceptional talent going into science and those going into other areas of study. All the data seems to indicate a tremendous stability in underlying interest patterns. People do go from physics to biology and back again, but there is no fundamental pull between science and the rest of the scholarly world.

TEACHING AND RESEARCH

Question: Has the teaching process suffered by the virtue of the overemphasis on research as the criteria in evaluating faculty members?

Mr. Halford: I would say not. I think the sharp division which the university has to make, and which Columbia is attempting to make, is between activities that involve students and those that do not. I would add that emphasis on research is not a peculiarity of science or government-supported research. I find that it is just as prominent a consideration in the classics where nobody will give a penny for the support of scholarly work.

Comment: I am bothered by the extent to which we are developing universities which are research universities, increasingly derelict in their teaching responsibility. Most of the research money is allocated to just a few prestige institutions. The same tendency to allocate funds to prestige institutions is true of the large foundations. Not only does this raise questions about the allocation of research funds, but it also raises questions about the allocation of teaching responsibilities. Research fellows who receive $7,000 or $8,000 a year are not easily recruited as faculty members by institutions which can offer no more than that for teaching jobs.

Mr. Halford: I am satisfied that there hasn't been any withdrawal from teaching responsibilities on Columbia's part. The university operates with two different budgets for two different purposes. We have university income from endowment and student fees, and we use these to recruit a force for teaching students. We then seek the help of foundations and the government to enable that force to do its scholarly work better, but we don't excuse them in any substantial way from teaching. Our people are teaching just as much today as they were thirty years ago.

Comment: You ought to insist on a significant surcharge on research for the purpose of expanding your teaching activities.

Mr. Halford: Unfortunately we are a nonprofit organization and the General Accounting Office of the U.S. government makes it a rule that we can be reimbursed only for the actual costs of research.

Comment: But perhaps a group of universities could force a change in this government policy. One by-product of this might be to permit other institutions to do more research and you to do more teaching.

Mr. Halford: If suddenly money to support research is made available, it will initially go where the talented people who can do the research are. There are certain institutions where these people are concentrated. I think, however, that we are reaching a point here, at least in some areas, where we can detect that these institutions have approached saturation. Then we will see this initial rapid rise in federal research funds given to these few institutions come to a plateau, followed perhaps by a gentle rise in activities everywhere. The solution to the problem of geographic distribution of research activities is going to follow as a second phase of development.

We have been through an explosive expansion of research funds. Five years ago we at Columbia had just completed a year with $12 million of federally supported research. I predicted then that by now if we were lucky we might get $20 million. Instead it is over $44 million. This money has gone to the people who were able to use it. I think that in the coming years we will see a much slower rate of growth. Columbia will remain relatively static. Many other places will begin to show rising support for their work.

Question: I came to Columbia in 1939 to join one of the scientific departments. The department has grown, but we teach just as much as we taught in 1939, and we are producing scientific papers at the same rate as in 1939. Why is there this distrust of money going to support scientific research? The money we get goes partially to the general funds of the university and therefore may help any other department. Does anyone think that medieval poetry is not being taught as well today as thirty years ago because we have new laboratories?

Comment: If the federal research funds are reduced then some universities and colleges may have to opt out of the physical sciences almost completely, just as there are very few countries that can stay in the arms race. This may mean that the number of institutions where original work in the sciences can be done may get smaller and smaller.

A lot of the money for research is a hidden subsidy to education in that it goes to the salaries of graduate and post-doctoral students, but I wonder whether an element of confusion may not arise through using the research grant mechanism as a substitute for straight educational subsidies.

Question: The whole emphasis here has been that Columbia could survive if the government ceased to support research, but it is clear that the government cannot afford to stop this support. The real question is, when do we set up a universities' grants approach, in which the government gives general support to higher education so that the universities can fulfill all their functions, without this distinction between research and teaching?

Mr. Halford: The leadership in this matter could come from the National Academy of Sciences committee which recently issued a report on federal support of basic research in institutions of higher learning. Except for differences of style and language, there is a complete agreement between their views and what I have wanted to express. I am, however, disappointed that they did not go further. Their criticism of the project-grant type of support is rather weak.

Comment: The National Institute of Health and the National Science Foundation have been making small block grants for several years. But what will happen when 1,500 institutions all around the country, each with its own representatives, try to get a share of a block grant? It will probably be an arithmetical division. In our kind of political structure, short of much greater sophistication, block grants simply means division by states.

Question: In its larger context, the tremendous expansion of research within the universities has had a fundamental impact upon American education. Some dozen or two dozen major universities have been pulled up to a level as research institutions that is well above the second level of universities. What does the creation of two dozen elite research institutions do to American education as a system? What does it mean for the future of higher education in the United States? I don't think that administrators have really faced up to these questions. We have a mechanism here for pulling scientific advance rapidly ahead. It will also hasten the creation of a technological society which is going to demand more from all of us. We are going to have to upgrade the education and the competence of millions of people.

Yet we are starving the lower educational levels where the people will have to be trained to take an effective place in the new
society that is being built partly as the result of what the universities are doing. I am concerned that there has been practically
no research on these problems.

Mr. Halford: There has always been an elite of a couple
dozen institutions that produced research in this country. We
have made them visible at the moment by pouring a certain
amount of money into them, but new knowledge has always
been contributed by a small number of institutions.

Question: The Space Agency and the Peace Corps both
granted about the same amount of money to Columbia, but are
those grants comparable? Is the Peace Corps money largely for
operations or service which are strictly speaking neither research
nor teaching?

Mr. Halford: I asked the dean at the School of Social Work
why the School is carrying out a program for the Peace Corps,
because the money primarily is spent in training volunteers to
go to Colombia in South America. We give these people some
field experience in social situations in New York. The dean said
that we are learning through experience that our approach to
education in social work has been seriously wrong. We have
taught people theory for several years before we have exposed
them to an actual problem. Because the Peace Corps requires us
to expose people to problems before they have any theory we
have found out that their students are actually performing
better than ours. Is this research in education or is it a training program? You will have to draw your own conclusions.
Moreover most of the National Aeronautics and Space Administration money comes to Columbia in support of training programs, although the content of the training programs is nowhere near as specific as is that of the Peace Corps. But still the
Space Agency's investment on this small scale here is largely to
train more space scientists.

THE ALLOCATION OF RESEARCH FUNDS

Question: Are there good projects in certain areas that are not being supported because it is not currently fashionable to do research in these areas while in other areas work perhaps of not equal caliber is being done because money is available?

Mr. Halford: If I can adopt a definition of "equal caliber" as being simply peer judgment in different fields, then there are very gross imbalances in the support available for the sciences. Chemistry departments are big because the subject is heavily required in undergraduate education. But there is only $4 million in the National Science Foundation which supports some of this activity. There are probably something on the order of four hundred persons competing for that $4 million. On the other hand the Atomic Energy Commission can furnish between $1 million and $2 million at the Nevis Laboratories [1] alone to support ten faculty members and fifty students. There are areas that are relatively starved for funds. Good projects are not necessarily being turned down, but the researcher may need $20,000 to carry out his research, and perhaps he gets $6,000. This may be almost wasted money in some instances because the researcher can't do with $6,000 nearly as much proportionally as he could do with $10,000 or $15,000.

Question: Are there some people who could do $6,000 worth of research but have to use $30,000 because that is the way money is being granted?

Mr. Halford: On the whole I think the situation is pretty well under control because most of these projects are subject to peer group judgment and peers are in competition for these funds.

Question: Is it more difficult today to get a new proposal initiated than it would have been ten years ago?

[1] The Nevis Laboratories are located at Irvington-on-Hudson, N.Y. A central facility of the laboratories is a cyclotron.

Mr. Halford: We had much stricter procedures in this university originally than we have now. In the case of certain agencies we do almost no monitoring of new proposals now, because we know that the agency to which the proposal is submitted can do a better job than we can.

Comment: I think there may be serious objections to administrative arrangements where one looks carefully at original proposals, but where the renewal process in some sense is easier than the initiation process—an arrangement which the government, I gather, also follows. This may be all right in an expanding economy, but in science perhaps one should be quite free in initiating inquiries but then be more and more careful about automatically renewing them. I gather too that the mechanism for cutting off grants to a man who is in the habit of getting grants is not too well developed.

Question: Are there any statistics about the number of proposals that are judged good by their peers but are not permitted to go through?

Mr. Halford: When a proposal is originated in the university it is by a faculty member. It is his idea. We will not accept it centrally unless it is endorsed by the department chairman. Most of the new proposals then go out to agencies like the National Science Foundation that circulate the proposal among a number of peers in other universities and solicit opinions. We have taken the position that a member of the faculty who is eligible and who wants to put his own professional reputation on the block nationally has a right to do it.

Question: I would like to know how many grants simply increase information which is of very little value, and how many really do increase knowledge. Is the grant-giving process in fact well-fitted to increase knowledge rather than information? I think this is the crucial issue.

Mr. Halford: I cannot make your neat distinction between information and knowledge. I know of no great scientific breakthrough, of a kind which I am sure you would recognize as an increase in knowledge, which was not based upon a tremendous mass of what you would call simple information.

Question: Are the projects that are getting through the imaginative and novel ones?

Mr. Halford: This is a peer group judgment and no one of us can answer for all the peer groups in the country.

Comment: Watching the process of peer judgment over four years I think one always wants to build in administrative controls against fashion. The National Institute for Mental Health has used a system of small grants of about $4,000 for any younger man who had a reasonable proposal. This is a direction in which one could move.

Comment: Before we agree to leaving this entirely as a matter of peer group judgment, it should be noted that it is also a matter of the way that Congress in particular feels about cancer or something else. It is a peer group judgment, but only within certain programs, and I think this makes a real difference.

Comment: If nobody knows where scientific discoveries are going to come from, it follows, I think, that the largest freedom and the greatest area of choice should be allowed to the scholars to pursue the research that they feel will be most likely to lead to breakthroughs of knowledge. At the same time we appear to be too satisfied with a pattern of research support which channels research money into certain areas more heavily than others for national purposes.

PRIVATE AND PUBLIC SOURCES OF RESEARCH FUNDS

Question: Do you notice any difference in the impact on the university of money given by foundations as compared with government money?

Mr. Halford: The transactions between the university and the private foundations are, from an administrative point of view, an uncharted jungle. If a member of any of the faculties of the university were to submit a formal proposal to any established government agency, it would be returned within forty-eight hours with a request from that agency for confirmation that the proposal had the official blessing of the university. This just doesn't happen with the foundations. We will get a letter with check attached saying here is the money for such-and-such a purpose, while we in the university administration did not know that any request had been made.

Question: If the university were free to distribute $44 million as it saw fit, how different would the distribution of funds be from what it is now?

Mr. Halford: The only answer I can give must be very indefinite and qualitative. I am sure that the distribution would not be identical with what it is at the present time, although it is interesting that federal funds are flowing in increasing amounts to the support of exotic languages, Chinese, Japanese, and the other language and area studies created under Title Six of the National Defense Education Act. The National Science Foundation has also been able gradually to enlarge the conception of its mission so that it will support a broad spectrum of activity.

Question: How does the thrust of the support given to the university by the private foundations compare to that of the government?

Mr. Halford: Usually private foundation grants apply over longer periods of time than government grants, although the federal support itself is underwritten for a longer period of time than one might realize. For example we carry on our books all Public Health Service money as expiring within a year. It does technically, except that the agency guarantees support for as long as five years, subject to congressional appropriation. It is my impression that most of the private foundation grants in recent years at Columbia have gone to the development of area studies. A few years ago we received a large grant to run for ten years to develop area studies in international affairs. The professional schools receive rather substantial grants, but it is not necessarily research money. It may be money to develop programs. There is a very generous grant from the Ford Foundation to help our School of Business develop a Ph.D. program through fellowships and other forms of support.

Question: During the past years there has been great pressure on corporations and other groups to provide unrestricted funds to the universities for the support of the graduate faculties as a principle source of teaching and new knowledge. At the same time you seem to imply that there is now a happy equilibrium between the available funds and the available manpower. Is there really this need for large unrestricted funds?

Mr. Halford: I have no difficulty in answering this question. The need is tremendous. There are all sorts of scholarly activities at Columbia that do not qualify for a penny of support from the sources we are talking about. In my role as a graduate dean instead of a research administrator, I would not hesitate to come right to your doorstep and ask for more money for dozens of fields like French literature or Sanskrit.

In spite of the fact that we have available for specific purposes the large sums of money that I mentioned, we have a very small margin of unrestricted money on which we operate the university today. The amount of money that we can actually control and put where we want to is so small in comparison to what we must do that we are getting into a tighter and tighter strait every year.

BIG SCIENCE AT THE UNIVERSITY

Question: You have said that you are not engaged in "big science" but that you have a lot of small cells of scientific activity. When enough small cells are put together, a gigantic organism results. What is your definition of "big science"? It seems to me that you are engaged in "big science," simply in terms of your budget, and in terms of the fact that your scientists are "big scientists," some of whom are engaged in decision to allocate funds for other "big science" projects.

Mr. Halford: If you have $45 million to support a program, all the elements of which are coordinated and related to each other in some planned way, you would surely have "big science." Or when you go down to a valley in West Virginia and see a radio telescope there which fits in nicely with the surrounding mountains and then realize that it is twenty-seven stories tall and that it takes a huge crew of people to design and build it, you have "big science" because you need an entire engineering organization that doesn't exist in the universities.

Question: Is the Nevis Laboratory an example of "little science"?

Mr. Halford: I think so even though there is a big cyclotron at the Nevis Laboratory. It is true there is a project here in the university, which I don't care to bring into prominence, which is "big science." It is an organization of several hundred people, and it exists for a particular purpose. I glossed over that by excluding the ten largest contracts held by the university. After you have excluded these you still have 75 percent of the total going to 99 percent of the projects, and there is absolutely no coordination or planning of any kind between any one of those and any other.

Comment: It seems to me the fact that you are coordinator of research grants would indicate that there is some measure of coordination.

Mr. Halford: Administrative control in this university is purely permissive. We don't initiate projects, we don't sell the services of our faculty. We have a very interesting control system. I cannot enter into a contract. I cannot commit the trustees of the university to any financial arrangement with anyone. The treasurer and the controller of the university alone have that power, but they will not exercise it without an instruction from me. They don't sell the services of the faculty for business reasons, and I can't enter into foolish business arrangements for good academic reasons.

RESEARCH FUNDS AND THE SOCIAL SCIENCES

Question: To what extent do you think that the relatively small amount of money that goes into social science research reflects the lack of imaginative proposals put forth by university social scientists? It seems to me that it is much more difficult to define a social science research proposal than one in the physical sciences without getting into controversial political and social areas. It also seems to me that much more money is available than is being obtained.

Mr. Halford: I agree with you. I think that the agencies which grant federal funds are very sensitive to becoming involved in controversial issues for fear that it may adversely affect not merely their support of social sciences but everything they might try to support. But I think there are other factors at work. For example, I think that it may be the relationship between the mentor and the student in the social sciences is quite different from that in the natural sciences. To polarize it completely, I think that it is difficult for my colleagues in the social sciences to use graduate students as assistants in their research activity. To the extent that there is any truth in that observation, it is a powerful deterrent to getting appropriations to support assisting activities and to extend the sphere of these activities. In the early days a very substantial part of the funds for the support of physical sciences in the universities was to increase the pool of trained manpower. There was a brief phase when Congress even accepted the idea of putting a floor under rather than a ceiling over the support of basic research in universities exclusively on that one point, to increase the supply of trained manpower. In the physical sciences the student is an apprentice and in a certain sense a very real equal in the research enterprise that is being supported. I don't think that there is the same type of collaboration in the social sciences.

Comment: Why is it that we social scientists are utterly unable to convey the need for our research? I think that social scientists must convince the American people that basic research must be carried out if we are to handle the manpower problems of the United States and attack sensibly the fundamental problems of poverty. Congress has passed the Manpower Training Act, which makes available a few million dollars of research money. It is, however, a very small amount compared to what is available in other fields and compared to the need for research.

Summation

by AARON W. WARNER

Professor of Economics, Columbia University, and Chairman, Columbia University Seminar on Technology and Social Change

THE BASIC ATTRACTION of the Seminar on Technology and Social Change lies in the opportunity it affords for a lively exchange of views. In most cases the presentations of the speakers are informal, and represent together with the discussions an unfolding and exploration of ideas rather than any statement of predetermined positions as the result of research. To a considerable extent, the views expressed reflect intuitive judgments based on a rich experience in activities which border on the problems discussed. The value of the proceedings is consequently to be found in the insights they afford on a vast range of subjects, and minimum attention has been paid to logical sequence or any precise ordering of topics.

The purpose of this chapter is to restate the substance of the discussion within a framework that hopefully will be more closely knit. The materials are drawn (and condensed) from the various papers and discussions, and their essentially intuitive character remains in evidence. However, in their totality they constitute a significant commentary on a well-defined set of problems which the seminar had undertaken to explore. These begin with a consideration of the complex interrelationships between science and technology, continue with an evaluation of the importance of science-based technology for the fulfillment of various social needs, and conclude with some observations concerning the conscious direction of science and technology in an effort to obtain optimum results from the huge expenditures that are entailed.

INTERACTION BETWEEN SCIENCE AND TECHNOLOGY

Scientific Advance as the Basis for the New Technology. The increasing importance of science as an underpinning for new technology is widely accepted as a fact. Technological innovations in earlier periods were largely the products of mechanical ingenuity rather than innovations based on science, and it is still true that in only a relatively few industries has the technology of the industry been linked directly to science, notably in the chemical, communications, power, and aeronautical industries. The more recent advances in science, however, portend a much greater impact of science on technology.

This impact can be described in several ways. One approach is that science itself is a source of future technology through the contribution of new knowledge, new components, and new techniques leading to new industries. In support of this view, one can point to the invention of new powerful scientific devices which have not yet found application, but which may determine important technologies of the future. Another approach regards basic science not as a direct source of new technology, but rather as a necessary (although not a sufficient) condition for the successful exploitation of technological ideas. On the one hand, the occurrence of scientific advance does not by itself assure technological exploitation; on the other hand, ideas for technological advance involving science may occur before science has developed to the stage where the new technology can be implemented.

We are thus faced with the question whether science inspires technology, or whether science is a limiting factor in self-motivating technological advance. Examples can undoubtedly be found to support both points of view. In either case, the advances in science present new vistas for an unparalleled expansion of technology.

An interesting observation is that new scientific devices that may lead to important breakthroughs in technology are rarely if ever developed by industrial companies, but are developed almost entirely by those who work in universities or in special laboratories. This presents the problem of how to bridge the gap between the scientific laboratory and the place where science is to be applied. A new phenomenon in this regard is the practice of scientists and engineers of getting together and forming small specialized firms, thus providing an institutional setting for practical applications of scientific discoveries.

As a further means of bridging the gap, it is suggested that public funds might be used to subsidize basic research in private industry as well as in government laboratories and nonprofit institutions. This is based on a belief that efficient production of technologically advanced products requires the close linkage of research, development, and production in the same organization.

Note must be taken of the importance of war as one of the great forces leading to the utilization of science in technology. The period since the war has seen a tremendous expansion of science. One result of the war was to provide money for scientific research, thereby enabling a whole new group to enter science. Many of our larger universities now derive a substantial part of their research funds from government sources. It is important to inquire, therefore, into the effect of government support on research and on the universities as teaching institutions. It is also of interest to ask what it is that motivates the government in supplying these funds. These inquiries take on additional significance in the light of the belief that the process leading to government support of science is irreversible and that there will never be a return to the haphazard support of science that existed prior to the war.

Scientists as Innovators. Are the people who make the discoveries in science also the innovators of new technology, or are different talents and characteristics needed for science and technology? Do scientific discovery and the application of science to new technology require different kinds of persons with different kinds of imagination, or is it mainly a matter of motivation, with the same qualities going into either pursuit?

There is no consensus on these questions. One view is that the process leading from scientific knowledge to technology and its applications can be divided into stages, with different kinds of persons and different types of imaginations operating at each stage. There would thus be separate categories for scientists, inventors, technologists, and engineers engaged in production. This view concedes that the same person can be both an important scientist and an important inventor, but points to the infrequency with which this occurs and to the fact that scientists who have made great scientific discoveries, and who have then become preoccupied with the application of their discoveries, rarely return to pure science. Another view holds that the same qualities of mind are involved at the various stages, but that there are differences in emotional attitudes. Another view is that when given the proper environment a scientist may simultaneously and successfully engage both in pure research and in the technical application of his findings.

The division of function between science and its applications raises additional questions in regard to the maintaining of a proper balance in terms of the people who are trained in the various categories. It is possible, for example, for a country to be preeminent in technology but backward in science, and the reverse situation may also occur. In recent decades, the United States has made tremendous strides in science. We have also produced an adequate supply of inventors, due to the existence of widespread higher education, and advances have been made in improving the quality of technicians engaged in translating pure science into technology. Generally speaking, our comparative advantage in technology appears to lie in organized effort rather than in individual genius. Individuality and genius, although they are not lacking in the United States, tend to flourish more in cultures in which there is greater stress on individualistic traits, as in France. We appear to have the greatest success in taking the grand inventions that originate elsewhere and developing them through teamwork.

Factors Conducive to Technological Innovation Originating in Science. Although by no means all-inclusive, a number of factors relevant to the stimulation of science-oriented technology may be identified.

There is clearly a need for a more widespread scientific literacy if technology based on science is to be encouraged. The danger must be avoided of creating a scientific elite whose property science will remain, and who will be unsuccessful in diffusing science into society as a whole. The conditions for bringing about such diffusion include the training of a sufficient number of scientists for both pure and applied work, and a greater scientific literacy in the community as a whole. This must include among other things the training of a large number of people with sufficient scientific sophistication to be successful inventors, the producing of a scientifically literate management with an appreciation of the potentialities that exist in the exploitation of scientific knowledge, and the preserving of mobility among intellectual disciplines, particularly in the movement of people from basic science into technology and industry. In those industries which have been most successful in producing technological innovation, an important element has been the movement of scientists into the industry and their gradual advancement up through the business hierarchy.

One suggested source for stimulating technological advance is the "overproduction" of scientists. This suggestion stems from a study of the chemical industry in Germany, which is said to owe its early development in the 1880s to the fact that there were more chemists trained than could find positions in the universities. Inability to find academic posts or their equivalent presumably can be an inducement for scientists to find careers in industry. There are, however, certain drawbacks to this approach. Unless circumstances permit their absorption by industry in suitable occupations, an overproduction of trained personnel may merely result in many individuals having to work at jobs below their scientific competence, with no significant increase in the number of scientific inventions or technological innovations. Any program for "overproducing" scientists must therefore be accompanied by a broader plan of economic development.

The question of a need for overproducing scientists may become academic in any case. In this country, the present shortage of scientists is in part a function of the slowing down of the birth rate in earlier decades. With the more recent increase in the birth rate, a larger number of persons will inevitably elect to study science. The shortage situation may thus be self-correcting within a relatively brief time. An interesting point to be observed here is that there has been little change since the war in the percentage of people going into science and engineering as compared to those going into the social sciences and the humanities, despite efforts to encourage a heightened interest in the sciences.

Evidence of the mobility of scientists into other fields is furnished by data showing that of those receiving their Ph.D.s in the basic sciences, only about 20 percent end up in the basic sciences. The remainder go into applied science or other fields. It is also observable that in industry, scientists move readily from research into operations. It has also been suggested that scientists who become "burned out" by rapid advances in their fields may then turn to technology.

Science, Technology, and Education. The task of training

scientists and technicians who will be capable of innovating in fields involving science focuses attention on our system of education.

If the potential of science is to be fully realized, education as a whole must become more scientifically oriented, not only for people who intend to become scientists, but also for those who will enter other fields. Education in science must become more selective, better organized, and more concerned with the context of what is taught and the attitudes that are inculcated. This is a far-ranging task which affects education at every level. One reason for the lack of diffusion of science may be the failure to reach back far enough into the educational process.

The major source of scientific training is the university. The question may be raised, however, whether research in technology is lagging because of snobbish attitudes toward applied science that are inculcated in the university. This could be attributed not only to the normal university environment which favors the study of pure theory, but to the additional factors of fashion, prestige, and the ease of securing government funds for "pure" research. There can be no objection to pure research; the university, however, has an additional responsibility to train students for a broader range of scientific activities in preparation for the varied types of jobs in industry and government, as well as in the universities themselves. At the same time, there is a danger in drawing too fine a line between pure and applied science, since one never knows what developments in pure science will have valuable applications. A related problem is the potential neglect of areas of study which lose fashion in the university community. Many of the areas of science which become less fashionable as fields of study nevertheless retain their basic importance and should continue to be taught.

The suitability of the departmentalized structure of the university for the teaching of science is also called into question. Modern applications of science require knowledge of many related subjects and frequently call for an integration of fields that transcends the narrow boundaries of the various departments. One manifestation of this need for integration of fields is the appearance of institutes at various universities, bringing together the different disciplines. What may be needed is a new division of the university—the "School of Advanced Study"—which will be concerned with interdepartmental work and will in addition provide a place where new disciplines can be brought into being and find a home.

Another range of questions relates to the impact upon the universities of federal support for science. To what extent have the universities become dependent upon federal support? To what extent have they become involved in "big science" to the detriment of other science, the humanities, and the social sciences? Has the teaching process been impaired by reason of the emphasis on research? Have the relatively few large university research centers been favored over the smaller colleges and universities in the granting of research funds?

The influence of government support is obviously great, although it can be argued that the effect upon the university is not as far-reaching as might be supposed. While government funds permit the broadening of research activities, they do not necessarily change the situation in regard to teaching, which (in the example cited in the discussion) continues to be supported from the university's general funds. If government funds were to be withdrawn—an unlikely prospect, however—the university would presumably find its teaching function unimpaired, although research would retrogress to the older methods. To protect themselves from the distorting effects of "big science," a number of universities have found it possible to join in setting up a new type of nonprofit corporation which acts as a buffer between the universities and the government in holding the government contracts for the operation of projects involving enormous facilities and large staffs. Another means of avoiding distortion within the university is the insistence upon the recovery from the government of all indirect costs related to the performance of research under government sponsorship, thus avoiding any diversion of university funds to provide hidden and inequitable subsidies to the sciences at the expense of other branches of the university. Finally, although it is clear that a disproportionate amount of the government funds has gone to the larger research centers, this may be reaching a point where the allocation of federal research funds to larger institutions has reached saturation. The initial rapid rise in federal research funds to a relatively few institutions may be followed by a second phase of better geographic distribution.

There is no consensus on many of these points, and a number of additional questions arise concerning the allocation and the use made by the universities of government funds. To the extent that federal support succeeds in channeling research into certain areas more heavily than others for "national" purposes, such as defense and space technology, it detracts from the ideal of permitting scholars to have the greatest freedom to pursue research that they feel will be most likely to lead to breakthroughs of knowledge. There are also moral problems involved in getting money for education and science under the guise of making a contribution to the defense effort when this is clearly not the intent of the researcher. It is also questionable whether the universities, in spite of the protective measures that have been described, have succeeded in avoiding distortions. Institutions which can supplement teaching stipends with research funds, for example, are in a better position to recruit for their faculties. There is also a feeling that there are gross imbalances in the support available for the sciences, and that there are other important research areas relatively starved for funds.

These and other difficulties lead to a consideration of the desirability of replacing the present method of government support by a universities grants approach. Under this method, the federal government would give general support to higher education so that the universities could fulfill all their functions without having to discriminate between research and teaching. Although this has its desirable features, the difficulty lies in the equitable distribution of the grants. Given the existing political structure, and in the absence of the unlikely adoption of a much more sophisticated approach, block grants to universities would simply mean division by states rather than by criteria related to educational needs and advantages.

Education at lower levels than the university presents another range of problems. On the one hand, it is recognized that the curriculum content in our schools must be responsive to advances in knowledge, and that every major advance in science requires a reconsideration of the preparation required for an understanding of the new developments. This is particularly true in the case of the preparation required for advanced training in science. In this regard, scholars must worry more about the implications of their scholarship for the whole of the educational process. On the other hand, it is felt that there are dangers in current trends which introduce abstract concepts into the high-school curricula. The fear has been expressed, for example, that experiments in the teaching of the "new" mathematics may lead to a generation of mathematical illiterates. There is also the possibility that specialization in science may become too narrow, with the result that young people will have little sympathy for science outside their own specialties and will be less able to carry out applied research. There is also the danger of paying too much attention to the "baroque" elements in mathematics and science. This refers to aspects of these fields which have relevance to special types of inquiry but may have little bearing on the more general and ordinary problems encountered in these disciplines. High-energy physics, for example, has "baroque" elements insofar as it deals with phenomena which are not encountered in ordinary matter under ordinary conditions and which bear little on the rest of science. There is no objection to the teaching of any kind of science, no matter how "baroque," so long as it does not demand too great a commitment of resources and so long as it does not impose an extreme viewpoint on the whole field of scientific and technical thinking. There must also be some process whereby those who do not want the "baroque" can be offered some other type of approach.

PROBLEMS OF CIVILIAN TECHNOLOGY

Civilian needs. The feeling was clearly expressed that the discussions should not be limited to how science can be used to improve technology, but should consider how an improved technology can be utilized to serve human beings. This, ideally, should include consideration of the appropriate rate of technological change and an awareness of its social cost.

Military and space activities in large measure determine the character of advanced technology in the United States. However, expenditures on these activities are based on considerations of national security rather than on the desire for scientific exploration, or on social or economic needs. There is also a vast difference in the approach to expenditures in these different sectors. The huge expenditures for military and space development are aimed at increasing the "technologically possible"; they are evaluated in terms of operational effectiveness as related to cost rather than in terms of social benefit. Moreover, they are not subject to the social, political, and economic constraints which customarily tend to set limits on expenditures for civilian needs.

The efforts of the private sector in research and development have lagged behind those of the military. To some extent, this may have reflected a belief that investment in research and development for military needs and space exploration would produce technological developments which could then be exploited by private industry in civilian markets. This may have some validity for the long run, but has less application to the more immediate future. There can be no doubt that the science supported for the military and space effort—for example, nuclear physics and solid-state physics—will have a long-term effect on our civilian industrial capabilities, but it is a costly, roundabout way to take care of our civilian needs. Important civilian objectives, such as education and the development of our vital civilian industries, should be supported explicitly and on their own merits.

The most important civilian needs to be met in the immediate future through the use of science and technology may be tentatively listed as follows: first, the creation of new jobs, in part through the development of new products and new markets; second, the rebuilding of our cities, which offers the single most important business opportunity in the country; third, the extension of education, where we must learn to apply the new techniques that have already been developed for handling and disseminating information; fourth, the utilization of our advancing technology in decision-making at the city, state, and national levels, where it should be possible to apply systems analysis to complicated economic and social problems; and fifth, the finding of a way to disarmament through the use of technology for arms inspection, and thereby releasing people previously engaged in defense activity for the task of solving the complex problems of our urban society. These tasks require a group of people with a deep understanding of economics, politics, and the social sciences, as well as of science and its applications.

To meet these evolving civilian needs, we must face the vital problem of reducing our military programs. Resistance to cutbacks in this area arise because of fear of the economic consequences. However, situations of this kind have arisen in the past and economic solutions have been devised. There is no reason to doubt that solutions can again be found. Problems of a special nature relating to science and technology that will arise from the conversion of technology from military to civilian uses include the retraining and reallocation of scientific and technical personnel to civilian tasks, in some cases requiring their return to sales and manufacturing activities at lower salaries, and the readjustment of former defense-oriented industries to market conditions where costs play an important role in decisions.

Introducing Technology into Backward Sectors of the Economy. In many cases, the limitations to the application of science and technology in civilian areas are not so much technical as social. Thus, in the housing and construction industries, where systems analysis, computer techniques, operations research, and new methods of construction could make adequate and cheaper housing more broadly available, advance is limited not by inability to apply science or by any dearth of technical possibilities, but by social, political, and economic considerations. Other industries to which these considerations would apply include the merchant marine and the railroads.

In these situations the problem is one of dealing with technical and social factors together. More generally, we must learn to use our more advanced technology to deal with the complexities of modern society. Although our problems are getting more complex, the power of science is also growing. Increasingly we are producing machines which can deal with complex systems and which can handle the vast amount of data that must be considered in making plans for a city rather than a house, for a transportation network rather than a road.

Do we have enough social knowledge to apply these new technical tools to social uses? One point of view is that we need a better understanding of people and institutions before this would be possible. The inability of our social institutions to keep pace with technological change would set a brake on the utilization of the new technology. To some, this suggests the need for social research to discover methods which would enable social institutions to change more rapidly. Another view is that the difficulty lies not in the inflexibility of our social institutions, but in the lack of an adequate effort to apply the new techniques. The obstacle is not so much the lack of knowledge, but rather the need to get greater commitment of people to do something about the social problems.

A special problem exists in the case of fragmented industries, such as the textile, coal, and housing industries, which cannot support the research and development necessary to keep them competitive. A dramatic illustration of the impact of research in this connection is the rapidity with which atomic energy has overtaken coal as an economical source of electric power. The fragmented industries tend to lag behind the integrated industries, as well as behind the industries in which research is government supported. It is argued that for such industries, government should take the responsibility for the relevant research and development and in this way help to overcome the technological lag. The obvious bars to effective research in the fragmented industries are insufficient funds, lack of tradition in research, and incompetent management.

One means of guiding technology into civilian uses would be to make more of the by-products of our military research and development available for private exploitation. The number of such by-products tends to be exaggerated, but there are nevertheless some important instances of such diffusion. An example of what can be accomplished along these lines is the establishment at the Oak Ridge National Laboratory of an Office of Industrial Cooperation, which calls the attention of interested entrepreneurs to developments at the government laboratories that may have commercial value. Another suggestion is that the government should use its leverage as a large buyer of civilian products to encourage technological innovation on the part of its suppliers.

The Importance of Entrepreneurial Skills. A consideration of another kind is the reminder that technology operates in the market place, and has no viability unless it can survive there.

Under modern conditions, the concept of a market must be given a broad interpretation. It implies not merely individual private buyers and sellers, but also the participation of government and various other collective groups. An imaginative probing of the market can be a crucial factor in the innovation process; if the market is probed in too narrow a way, potential demand for innovations will be underestimated. On the other hand, it may not pay to force an innovation if the market is not yet ready to pay for it.

There is a dearth of people who understand the business aspects of a technological enterprise, particularly the elements of risk-taking and financing. Innovation requires not only the inventor who conceives a product or service to meet a customer's needs, but also the entrepreneur who underwrites and markets the new product. In addition to finding a market for the new product, it may also be necessary to invent a method to sell it, and sometimes even to invent the customer.

The extent of entrepreneurial risk in regard to introducing technological innovation is not always appreciated. Historically, major new technologies have received social support in one form or another, as in the building of the railroads and the introduction of aeronautical technology. Major technological changes come about through "invasion" of the market, and society must stimulate "invasion" if it wishes to stimulate innovation. This requires the active encouragement of venture capital and a willingness by society to share in the risks. There may also be a need for some system of incentives to encourage entrepreneurs to put new products on the market after an innovation has taken place which would render obsolete their existing products.

Another range of questions relates to inadequacies in "traditional" American industrial management which make it difficult for industries to take full advantage of modern technology. In some cases, managements may have adopted the trappings of scientific method but fall short of having a scientific approach toward scientific inquiry and a willingness to adjust their methods to keep abreast of new discoveries. In other instances, a lack of technically trained people in positions of responsibility results in failure to deal in a competent manner with the complexities of modern technological developments. Another factor is the lack of patience of nontechnical managements in exploring the value of a new invention, and a tendency to look for immediate "fall-out" products from research that will bring profits, with no appreciation of the actual time span that would be required. One remedy for these defects is to place technically competent persons in senior management positions. But it is also necessary to inculcate more realistic attitudes toward the need for scientific research as opposed to methods of "simulation" which attempt to solve problems without first obtaining the basic knowledge that is necessary for success.

THE DIRECTION OF SCIENCE AND TECHNOLOGY

The Federal Government and Scientific Priorities. The importance of this issue is indicated by the fact that the government provides approximately two-thirds of the funds that are available in the United States for research and development. The decisions taken by the government in the allocation of these funds may thus be assumed to have a profound influence on the development of science and technology. The criteria used by the government to determine priorities, however, are vague, and experience indicates that the task of deciding where to spend huge sums on scientific research is fraught with uncertainties.

There is no objection in principle in a democratic society to political control of the allocation of government funds to science, in the light of their importance to national goals. The difficulty is that, having decided upon a goal, the relationship of the activity which is being financed to the desired goal is often unclear. No one really knows in advance which science will be relevant to any particular goal. The decision to support scientific research, therefore, must be based on other considerations.

One approach would be the view that science should be pursued simply for intellectual reasons, bearing in mind that the incidental gains for technology are likely to be substantial. This would still entail a judgment as to whether one intellectual opportunity is greater than another. A more sophisticated approach along these lines would be to consider both the magnitude of the intellectual opportunity and its cost, and to strike a balance based on these considerations. But a method for evaluating the cost effectiveness of scientific research has not yet been perfected. There are other difficulties as well. Even if a suitable measure could be devised, reliance on cost effectiveness may neglect important factors which are intuitive rather than quantifiable. Moreover, if science and its applications are regarded as central to the future development of the economy and to the maintenance of the nation's leadership in the world, the cost in relation to these goals may be inconsequential.

Another suggestion is that the relevant variables in making a choice should be intellectual opportunity and social need. This implies some method of maintaining a balance between opportunity and need that will achieve an optimum result. Following this approach, spending on intellectual opportunity would be tempered by what is socially needed; at the same time, spending on what is socially needed would be tempered by the state of the art. Behind this suggestion is the realization that it can be wasteful to spend money for projects for which the scientific underpinning is not yet ready, and that it can be equally wasteful to delay taking advantage of scientific and technical knowledge that is already available.

In the absence of a more precise measure, the only workable mechanism for arriving at decisions of this type appears to be reliance on expert opinion and argumentation by people who see the problem from different points of view. This method is actually the one that is used. The President's Science Advisory Committee has resisted being put in a position of making the decision as to specific priorities, and priorities are still in fact being determined by the Bureau of the Budget after argumentation and with advice from the scientific community.

Within the scientific community itself, there is a sense of what is important in terms of research as opposed to what would be of lesser value. This is not always a reliable guide, but it is the best available. The question is one of how to utilize this perspective in giving leadership to scientific endeavor. If approached with proper sensitivity, scientific activity could probably be successfully channeled. However, any organized effort to do this is not likely to succeed. The major difficulty here is that the persons who rise to positions of leadership in organizations are generally not the natural leaders of science, and this is particularly true of organizations that are strongly hierarchical, as in the Soviet Union. A more successful approach would be to stress motivation, such as exists in the Nobel Prizes, and to recognize the tendency among scientists to emulate the natural leaders in science. A similar type of problem exists in the selection of scientists to make policy recommendations to the government. It is difficult to recruit persons who are able scientists and who also possess the ability to assess the impact of current trends and other relevant considerations. Many brilliant scientists would simply be impatient with this type of activity. In both cases, the problem is one of finding and using the proper incentives and talent.

Another aspect of the government's problem in regard to expenditures on science is the role of Congress. On the one hand, there is increasing uneasiness among members of Congress because they feel they must rely too much on the executive branch for expertise, and because they have to take too much on faith when they vote large sums for scientific research. This uneasiness has manifested itself in the recent creation of congressional bodies to investigate government research activities. On the other hand, the practice of spreading authority over science programs among a large number of congressional committees makes it extremely difficult for individual members of Congress to have any coordinated view of the science problem as a whole.

This is further complicated by political considerations. With the large defense budget, congressmen are inevitably concerned with the amount of defense business they can obtain for their own constituencies. Since defense contracts are assumed to follow a path determined by the location of the great research centers and highly trained scientific and technical personnel, congressmen from districts in which these facilities do not exist are under pressure to lobby for the creation of new research facilities. So far as the geographical impact of research expenditure is concerned, congressmen are not too troubled over the exact scientific purpose for which the money is to be spent. Even if there is doubt that basic research in the universities or similar institutions tends to attract industry to the area, political considerations still make it imperative to put in installations that people can see. Considerations of this kind render impracticable the establishment of a congressional office of science and technology, since in any office serving the whole Congress the loyalties would be too diffuse to satisfy the needs of the individual members. An acceptable method for advising individual congressmen on science problems remains to be devised.

Industrial Responsibility for the Direction of Science and Technology. Another element to be considered in assessing the government's role is that over time research projects grow larger, take longer to execute, and become more costly. This increases the urgency of knowing how to select the right project, since it becomes more and more difficult to change direction once a project has been started. In dealing with this problem, the emphasis has been on the need for improving the methods used by the federal government in allocating funds for research. The question is now raised whether the government should be the major decision-maker in this area, and whether private industry is not in a more advantageous position to determine how research funds should be spent.

A number of reasons are adduced in support of a more active research role for industry. Increasing centralization of decisions concerning research and development is said to be inadvisable, since no one set of people is wise enough to forecast the most promising directions to be taken. In providing multiple centers for deciding upon projects, industry would thus provide a more reliable basis for judging the potential value of research. Moreover, it is claimed, industry would be less handicapped than government because of political considerations. The government, for example, must operate on the basis of a four-year cycle, which makes it difficult to manage projects of several years duration. This is clearly not an impediment for industry. It is also difficult for the government to foster a line of research that has scientific validity but is currently unpopular. Here again the government of necessity operates under more severe restrictions than industry.

These arguments provide an additional basis for advocating the subsidization of basic research in private industry. Since industrial firms participate in supplying the tax money which is used in federally supported research-and-development programs, the utilization of a limited portion of these funds for similar activities by industrial firms should not be difficult to justify. Following this approach, an ingenious method for inducing an increase in private expenditure on research and development with no additional cost to the federal government would be to reduce federal expenditures on research and development each year by a stated amount and to offer this amount to industry in the form of tax reductions for those firms who would agree to carry on research and development on a private basis.

These proposals in turn, however, raise a number of difficult questions. To what extent would large firms with established markets be disposed to undertake research and development in new fields? Unless such firms were already enormously impressed by the possibility of a transformation of the market, would tax benefits accomplish this objective? Must it be assumed that major innovations through research and development require the resources of large institutions? If so, how shall we deal with the situation in which new ideas originate in small companies which lack the resources to develop them? What can be done to provide greater resources and incentive to the small companies to undertake basic research?

The Responsibility of Scientists for Social Innovation. A final range of problems involving the direction of science and technology relates to the crucial role of the scientists in modern-day society.

It is recognized that the intellectual energy inducing change in our society stems from a fairly small community of scientists and related persons whose recommendations vitally affect our social, economic, and political structure. Yet these scientists are in no way attached to any of the traditional means of introducing social change, and their recommendations are customarily made without reference to their implications for society as a whole. To the extent that their recommendations pertain to purely scientific matters, this does not present any problem. But to what extent should we expect scientists to take account of the social implications of their activities? In this regard, it may be claimed that the scientific community has chosen to abrogate its responsibility. This, however, may be inevitable. Those scientists who take time to concern themselves with social consequences tend to be the administrators, while the working scientists appear not to have the time. Moreover, the fault is not that of the scientists alone, but of other intellectuals as well.

This abbreviated summary chapter obviously cannot do justice to the variety and depth of the views set forth in the papers and discussions themselves. However, an overall view of the proceedings may add perspective, and perhaps lead to a desire to pursue further some of the ideas presented. The questions raised may also suggest new lines of inquiry. The seminar will continue to explore these and related problems in the sessions to come and will undoubtedly succeed in probing more deeply into the complex issues surrounding the advance of technology and its implications for social change.

Participants in the Seminar, 1963-1964

R. CHRISTIAN ANDERSON
Brookhaven National Laboratory

CARL BARNES
Food Machinery Corporation

ARNOLD BEICHMAN
Electrical Union World

DANIEL BELL
Department of Sociology
Columbia University

MYRON BLOY, Jr.
The Protestant Ministry
Massachusetts Institute of Technology

CHARLES R. BOWEN
International Business Machines
  Corporation

PETER B. BUCK
National Academy of Sciences

A. C. BURSTEIN
Department of Commerce
City of New York

PETER J. CAWS
Carnegie Corporation of New York

NEIL W. CHAMBERLAIN
Department of Economics
Yale University

EDWARD T. CHASE
Cunningham and Walsh

EWAN CLAGUE
Bureau of Labor Statistics

DONALD COOK
Basic Systems

THOMAS E. COONEY, Jr.
Science and Engineering Division
Ford Foundation

CHARLES DeCARLO
International Business Machines
  Corporation

ALFRED S. EICHNER
Department of Economics
Columbia University

LUTHER H. EVANS
International and Legal Collections
Columbia University

VICTOR R. FUCHS
National Bureau of Economic
  Research

JOSEPH W. GARBARINO
Department of Economics
University of California

EDWIN GEE
E. I. duPont de Nemours and
  Company

ELI GINZBERG
Graduate School of Business
Columbia University

SILVIA B. GOTTLIEB
Bureau of Labor Statistics
U.S. Department of Labor

GEORGE A. GRAHAM
Brookings Institution

DANIEL GREENBERG
Department of History
Columbia University

ROBERT H. GUEST
Tufts School of Business Administration
Dartmouth College

MARY ALICE HILTON
Institute for Cybercultural Research

RONALD E. JABLONSKI
School of Business
University of Michigan

EARL D. JOHNSON
Delta Airlines

DAVID KAPLAN
Economics of Distribution Foundation

NORMAN KAPLAN
Department of Sociology
University of Pennsylvania

MELVIN KRANZBERG
Department of History
Case Institute of Technology

JAMES W. KUHN
Graduate School of Business
Columbia University

NORMAN KURLAND
New York State Department of Education

ERIC LARRABEE
Consultant

LEWIS LORWIN
Consultant

JOHN McCOLLUM
U.S. Department of Health, Education, and Welfare

DONALD N. MICHAEL
Institute for Policy Studies

PAUL NORGREN
Consultant

GEORGE W. PETRIE
International Business Machines Corporation

EMMANUEL PIORE
International Business Machines Corporation

GEORGE REYNOLDS
Department of Physics
Princeton University

ORMSBEE W. ROBINSON
International Business Machines Corporation

VIRGIL ROGERS
Syracuse University

JERRY M. ROSENBERG
Teachers College
Columbia University

MARIO G. SALVADORI
School of Engineering
Columbia University

DONALD F. SHAUGHNESSY
University Seminars
Columbia University

DAVID SIDORSKY
Department of Philosophy
Columbia University

BARRIE SIMMONS
Basic Systems

ARTHUR L. SINGER
Carnegie Corporation of New York

HAROLD J. SZOLD
Lehman Brothers

FRANK TANNENBAUM
University Seminars
Columbia University

JACOB TAUBES
Department of Religion
Columbia University

ROBERT G. THEOBALD
Consultant

MARKO TURITZ
California-Texas Oil Corporation

HENRY H. VILLARD
Department of Economics
The City College of New York

AARON W. WARNER
Department of Economics
Columbia University

E. KIRBY WARREN
Graduate School of Business
Columbia University

PAUL WEINSTEIN
Department of Economics
Columbia University

LAURENCE K. WILLIAMS
School of Industrial Relations
Cornell University

SEYMOUR L. WOLFBEIN
Office of Manpower, Automation, and
  Training
U.S. Department of Labor

CHRISTOPHER WRIGHT
Council for Atomic Age Studies
Columbia University